PROGRESS IN AGRICULTURAL GEOGRAPHY

CROOM HELM PROGRESS IN GEOGRAPHY SERIES
Edited by Michael Pacione, University of Strathclyde,
Glasgow

Progress in Urban Geography
Edited by Michael Pacione

Progress in Rural Geography
Edited by Michael Pacione

Progress in Political Geography
Edited by Michael Pacione

Progress in Industrial Geography
Edited by Michael Pacione

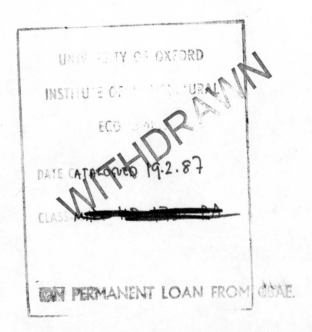

Progress in Agricultural Geography

Edited by MICHAEL PACIONE

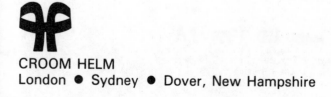

CROOM HELM
London • Sydney • Dover, New Hampshire

© 1986 Michael Pacione
Croom Helm Ltd, Provident House, Burrell Row,
Beckenham, Kent BR3 1AT
Croom Helm Australia Pty Ltd, Suite 4, 6th Floor,
64–76 Kippax Street, Surry Hills, NSW 2010, Australia

British Library Cataloguing in Publication Data

Progress in agricultural geography
   1. Agricultural geography
  I. Pacione, Michael
   338.1′09   S439

   ISBN 0–7099–2095–4

Croom Helm, 51 Washington Street, Dover, New Hampshire, 03820 USA

Library of Congress Cataloging in Publication Data
Main entry under title:

Progress in agricultural geography.

   (Croom Helm progress in geography series)
   Includes index.
   Contents: theory and methodology in agricultural
geography / B.W. Ilbery—classification of
agricultural systems / J.W. Aitchison—diffusion of
agricultural innovations / G. Clark—(etc.)
   1. Agricultural geography—addresses, essays,
lectures. I. Pacione, Michael. II. Series.
S494.5.G46P76   1986   630.9   85-31389
ISBN 0-7099-2095-4

Filmset by Mayhew Typesetting, Bristol, England
Printed and bound in Great Britain
by Billing & Sons Limited, Worcester.

CONTENTS

FIGURES

Figures

TABLES

Tables

To Christine, Michael John and Emma Victoria

PREFACE

Agriculture can be defined as management of the land for purposes of producing plant and animal products to satisfy human needs. It is both a way of life and an economic activity. Traditionally, the geographical study of agriculture was regarded as a sub-branch of cultural geography or economic geography, with strong links to cognate subject areas in economics and sociology. More recently, agricultural geography has emerged as a multi-faceted field of inquiry in its own right.

Study of the structure and pattern of agricultural activity has long formed a focus for researchers interested in agricultural typology, productivity evaluation, innovation diffusion and location theory and modelling. The techniques employed in the analysis of these topics have generally evolved with the subject to ensure that such themes continue to represent foci for investigation. Since the early 1970s, however, these traditional concerns have been augmented by new issues including the industrialisation of farming and its impact upon rural economic and social structures, land use change and conflict with particular attention devoted to the loss of good quality agricultural land to urban development, the institutional determinants of agricultural activity, part-time farming and multiple job-holding, agricultural marketing and distribution, land ownership and the agricultural land market, and the world food problem.

This collection of original essays is designed to encapsulate the major themes and recent developments in a number of areas of central importance in agricultural geography. The volume is a response to the need for a text which reviews the progress and current state of the subject and which provides a reference point for future developments in agricultural geography.

Michael Pacione,
University of Strathclyde,
Glasgow.

INTRODUCTION

Just as the history of Mankind is closely related to the history of agriculture, so agricultural geography has long been a key component of human geography. In Chapter 1 Brian Ilbery underlines the dynamic nature of the modern subject in his discussion of recent developments in theory and methodology. Three major theoretical perspectives (the environmental, economic and behavioural) are identified and the characteristics of each together with the principal analytical methods employed are examined. Although Man's ability to influence his natural environment has diminished the relevance of the classical deterministic interpretation, environmental influences on the spatial structure of agriculture (as in the concepts of ecological optimum and the margin of cultivation) remain of particular significance at the regional level. In addition the recent recession and the price-cost squeeze in agriculture have underlined the importance of yields and the intensity of production and have encouraged farmers to seek the most suitable physical areas for their enterprises. The related issue of energy-efficiency in agriculture has been prompted by the dependence on fossil inputs, while a further environment-related concern refers to the conservation of agricultural resources. The second major theoretical perspective seeks to understand the role of economic factors in the determination of agricultural patterns. While distance and transport costs were key elements of earlier models more recent formulations have emphasised forces such as the rate of urban development (as in Sinclair's model of urban fringe agriculture). Other factors which have disrupted the classical operation of market forces include the existence of agricultural cooperatives, farm gate sales and government intervention. Three aspects of the industrialisation of modern agriculture which have received particular attention are (1) economies of scale and increasing farm size, (2) enterprise and regional specialisation, and (3) the relationship between farmers and food processing industries. The failure of normative economic models to fully explain observed agricultural patterns led to the introduction of the behavioural perspective during the 1970s. This recognises the importance of social and psychological factors in agricultural decision-making. Techniques employed within this framework include models such as game theory which acknowledges that farmers make choices within an environment of uncertainty; innovation diffusion and the logistic curve of adopters;

1

and various strategies to analyse farmers' goals and values, including repertory grid procedures and point score analysis. In spatial terms a particular focus of attention is the dynamic farming environment of the rural-urban fringe. Each of these major theoretical perspectives has formed a focal point for research in agricultural geography during the post-war era. No single viewpoint in isolation is sufficient to explain contemporary agricultural patterns. What is required is an integrated approach which recognises the contribution each can make to understanding the complexity of the agricultural landscape.

Classification is a fundamental component of scientific enquiry and within agricultural geography the classification and regionalisation of farming systems represents a long-established objective. In Chapter 2 John Aitchison examines the practice and problems involved in the classification of agricultural systems. The discussion is structured around a four-stage framework which identifies the major phases and conceptual issues in the classificatory process. Each of these — identification of objectives, scale of analysis and data input, taxonomic methods and pattern interpretation — is then discussed in depth. Although some classifications have been employed to test the applicability of agricultural land use models the majority to date have been descriptive. While these exercises may be criticised for their low explanatory power they do simplify the complexity of the real world and can initiate more process-oriented investigations. It is suggested that the concern with broad typologies of farming types should be complemented by more policy-relevant studies of particular farming systems (such as a typology of marginal farms). Clearly, the scale of analysis employed will affect the nature of patterns produced. While studies based on the farm are of particular interest, this being the basic organisational unit of the agricultural landscape, data limitations often require recourse to more macro scales. The selection and operational definition of variables to be employed in the classificatory process are also of key importance. The difficulties of aggregated statistics, sampling procedures, levels of measurement and methods of standardisation are considered. It is suggested that despite conceptual and measurement difficulties the introduction of factors related to the behavioural environment deserves greater consideration. The variety of possible taxonomic procedures includes both assignment techniques (in which the researcher sets up an *a priori* group structure to which individual taxonomic units are then assigned) and classification methods (such as factor analytic procedures). The advantages and disadvantages of each type are examined. Methods of assignment including, for example, the least squares method, the use of ternary diagrams, and the

IGU classification of European agriculture, are considered to be more appropriate than classification methods for comparative investigations and for studies of temporal change. The principal advantage of the latter suite of algorithms (e.g. principal component analysis with or without a contiguity constraint) is their suitability for multivariate information sets and the fact that the resulting typology or regionalisation emerges from the data (i.e. is determined by the characteristics of the taxonomic units) and is not imposed. An illustration of how the two procedures can be combined is provided by a study of farming regions in France. In conclusion, it is suggested that insufficient consideration has been given to the question of the quality of particular typologies or regionalisation. It is recommended that this deficiency be offset by more rigorous examination of the effects of methodological differences on taxonomic results. In the final analysis, however, the utility of any classification is determined by how well it satisfies the objectives of the investigation.

The adoption of new techniques is an essential part of the agricultural development process. In Chapter 3 Gordon Clark seeks to identify the fundamental causes and effects of new farming practices. He identifies four main requirements for progress in the field: (1) a wider definition of an innovation, (2) re-assessment of the methods used to analyse spatial diffusion, (3) clarification of the relationship between innovation and economic development, and (4) closer attention to the social and economic situation of individuals. Contemporary research relating to each of these questions is then subjected to critical analysis. The difficulty of defining an innovation is discussed and the need for a clear identification of the population of potential adopters is emphasised. Alternative approaches to the study of innovation diffusion are then examined, ranging from the early cartographic studies to the seminal work of Hagerstrand which underlined the importance of the neighbourhood and hierarchical effects and the key importance of communication in the diffusion process. Given the growing volume and variety of information impinging upon both individual and corporate agricultural operators, testing the effectiveness of information flows represents a continuing challenge. Several models of the decision-making process are examined. An additional complicating factor is that the nature of the innovation, which can itself evolve over time as new models are produced as a result of consumer feedback, will also affect its rate of adoption. The fact that not all diffusion processes are demand-based suggests that for some types explanation must begin with analysis of the motives of the propagators. Attention then turns to the relationship between innovations and economic development, and theories of agricultural development in which innovation plays a key role are examined. The way in which this relationship can lead to regional and individual inequalities is also demonstrated. In order to gain insight into the complexity and, in some instances, apparent economic irrationality of decision-making it is necessary to complement generalised macro-scale studies and theoretical formulations with investigation of the impact of innovations on the social and economic position of individuals. Finally, three key areas of future research are identified: (1) fuller consideration of the processes which influence the kinds of innovation made available to potential adopters, (2) study of the reciprocal effects which operate between innovations and the structure of economies and societies, and (3) further investigation of how decisions on the transmission and adoption of innovations are taken by organisations and individuals.

In Chapter 4 Michael Troughton examines the dynamic character of agricultural systems in the modern world. Particular attention is focused on agricultural industrialisation, a process which involves a fundamental reorientation of farming from a small scale moderately capitalised activity to one in which the major part of production derives from a reduced number of large highly capitalised units. This phenomenon has been characterised as the third agricultural revolution. The process and characteristics of agricultural industrialisation are considered in a model which underlines the common nature and application of agro-technology across ideological boundaries. Four general processes are recognised, (1) increases in the size of production unit, (2) specialisation in production, (3) intensification of capital inputs, and (4) vertical integration of farm production with other parts of the agricultural-food system. The economic benefits of agricultural industrialisation can be offset by negative impacts which may be ecological (e.g. soil erosion, pollution, the narrowing genetic base of crops and livestock and concern over the nutritional quality of products), aesthetic (e.g. loss of hedgerows or intrusive building styles) or socio-economic (e.g. redundancy, rural out-migration and its consequences for communities and service centres, and regional disparities as a result of the concentration of economically viable agriculture into certain areas). Within the general framework of industrialisation, however, specific factors such as farm tenure, ownership of supply and processing facilities, and political ideology combine to produce varying models. Three main types are identified, (1) the pure socialist model in which the state controls all aspects of agricultural production, typified by countries of the Warsaw Pact, (2) the pure capitalist type based largely on private ownership of the means of production and the operation of market forces although with a degree of government intervention, as in North America, and (3) the co-operative type based on retention of individual ownership but use of co-operation to achieve economies of scale in purchasing, marketing and processing. This traditional involvement of farmer-based co-operative organisations has persisted in Scandinavia. While most attention has been given to the negative effects of agricultural industrialisation in the Developed World it is pointed out that in parts of the Third World, where elements of modern agriculture have long existed within the traditional economies, development of an export-oriented agricultural sector competing for resources with small-scale domestic producers represents a potentially serious economic problem. It is concluded that although the industrialisation process has affected only a monority of the world's agriculture its effects are pervasive and likely to be permanent since it emanates not from within agriculture but from the dominant industrial-urban sector which seeks to reduce the differences between primary production and the rest of the economy.

Most governments intervene in the agricultural sectors of their economies to some degree, with the level of involvement increasing along the socio-political spectrum from the capitalist economies of North America, through the redistributive welfare states in Western Europe to the centrally planned economies of Eastern Europe and the USSR. In Chapter 5 Ian Bowler examines government intervention in countries with developed market economies and democratic systems of government. The different values and sectional interests underlying the stated goals of agricultural policy are discussed first. It is pointed out that the agricultural lobby has exerted a disproportionate influence on policy-making through the efforts of farmers' groups and political voting power. Other interest groups which seek to influence agricultural policy include agribusinesses, food aid agencies, consumers and national interests within organisations like the EEC. Each country employs a distinctive set of measures to achieve policy goals and the factors that contribute to national policy formulation are discussed. These include, (a) stage of economic development, (b) farm size structure, (c) resource endowment, (d) population density, (e) degree of food self-sufficiency, (f) the political influence of agriculture, (g) the societal value system and (h) the nature of the farm problem, e.g. small farm inefficiencies, over-production, cost-price squeeze on producers, low farm incomes. The impact of intervention measures is then evaluated although the absence of specifically spatial objectives and of disaggregated data impedes geographic analysis of agricultural policy. Three particular questions are considered: (1) measuring the costs of agricultural protection. Both direct costs (e.g. deficiency payments, direct income supplements, grant aid for investment) and indirect costs (e.g. tariff and import levies) involve a redistribution of wealth within and between nations; (2) monitoring the income effects of government intervention. Since most measures aimed at maintaining farm incomes focus on product prices, producers with the greatest volume of output gain the most benefit. The alternative of providing direct income supplements has been resisted in most countries. The conceptual and statistical problems of making income comparisons are discussed; (3) assessing the efficiency of policy measures. Agricultural goals are pursued by manipulating farm inputs (e.g. capital is influenced by fiscal policy), outputs (e.g. via product price guidance) and farm structure (e.g. land reform measures). Five general conclusions on agricultural policy are offered. First, the policies of developed countries have had adverse effects on world trade, especially for developing countries. Second, policies have induced often unnecessary increases in farm production. Third, domestic product prices have been distorted in relation to import prices. Fourth, significant welfare and budgetary costs can result from protectionist policies. Fifth, the income objective is being achieved only for certain sections of the farm population. Finally, the need to examine non-market (i.e. political) reasons for the restricted achievements of policy measures is underlined. It is suggested that a new order of priority in policy goals is emerging with less emphasis on the narrow farm interest and more on the costs of intervention, the environmental consequences, and the effects on regional rural development and on the world food trade. This broader perspective is essential in order to understand fully the nature of government intervention in modern agriculture.

In Chapter 6 Andrew Dawson complements the analysis of government intervention in the West with an examination of agrarian reform in Eastern Europe in the post-war period. The discussion is based on a chronological structure which identifies four main phases in the evolution of East European agriculture, (1) agricultural structure prior to the Second World War, (2) the period of change 1945–60, (3) the period of consolidation 1960–80, and (4) the current situation. The structure of agriculture in Eastern Europe before the Second World War displayed considerable diversity ranging from small peasant holdings resulting from land reforms in the inter-war period (as in Bulgaria and Romania) to situations (e.g. in Albania and Hungary) of much greater concentration of land ownership, with in some areas (such as Western Poland) medium-scale commercial agriculture. Generally, the land reforms which occurred prior to the Second World War did not induce significant improvements in living standards or levels of economic development, partly as a result of world market conditions, population pressure and the lack of off-farm employment. The completion of pre-war land reforms was a central tenet of most of the East European states in the post-war era, and expropriation and redistribution of larger holdings led to the practical disappearance of privately-owned farms of more than 50 ha. and the proliferation of farms of about 5 ha. While this removed the pre-war problem of peasant indebtedness other problems created included a lack of infrastructure, unfamiliarity with newly acquired land, lack of skills and inefficiency of small holdings. Collectivisation accelerated during the 1950s with the rise of one-party communist governments and, with the exceptions of Yugoslavia and Poland, the process was largely completed by the early 1960s to produce an agricultural structure based on the Soviet model. The deviation of Poland and Yugoslavia from this model provides an opportunity to compare agricultural achievements under different systems, as well as to study the problems of centrally-planned economies with a large privately-owned agricultural component. During the period of consolidation from 1960 to 1980 state and collective farms were increased in size and state farm activities extended into food processing. Continued pressure was put on the remaining private farmers to co-operate with the system. Insight into the effects of these policies on East European agriculture is provided by an examination of the extreme positions represented by Albania, a model of Soviet-inspired state-controlled agriculture, and Poland, where private farms continue to operate. A general assessment of the achievements of agrarian reform in Eastern Europe would identify the lifting of the burden of peasant indebtedness, relief of rural population pressure, and the diversion of population and resources from the primary sector into extractive and manufacturing industries. The system, however, is not without its problems including shortage of inputs such as machinery and fertilisers, frequent failure to meet production targets and inter-sectoral competition between state ministries. Comparative evaluation of the agricultural systems of East and West represents a major research topic and more detailed analysis is required to provide a definitive answer to this question. This serves to underline the diversity of agricultural structures in Eastern Europe and the futility of attempting to construct a single model of East European agriculture.

The relationships between agriculture and urban development is one of the most contentious issues in contemporary agricultural geography and the urban fringe has been characterised as a battleground for competing land use interests. In Chapter 7 Chris Bryant defines the scope of the 'agriculture and urban development' field and identifies two fundamental sets of research questions. These centre upon (1) the process by which urban development has an impact on agriculture, and (2) the societal response to these impacts. Recent research in each area is then analysed. A fundamental distinction is made between direct (i.e. the removal of agricultural resources from production) and indirect impacts (i.e. effects on the continuing agricultural structure). Each set of impacts may be postitive or negative depending on whether it is viewed from the perspective of the individual or of society in general. It is suggested that while the research literature affords much attention to the negative effects of urban development on agriculture the evidence may be less clear-cut even in the case of direct impacts. Particularly important is the nature of the data employed and a discussion of census and map sources reveals possible ambiguities. Less detailed research is avialable on indirect impacts but one of the most significant areas concerns the effect of expectations of urban development on agriculture. A partial model linking investment planning and urban development is presented but further empirical evidence is required. In general studies of indirect impacts of urban development have set out to (a) draw inferences about impacts from macro-analysis of surrogate variables (e.g. the assumption that high land prices cause problems for farmers wishing to enlarge holdings), (b) focus on one specific type of impact (e.g. of urban development on agricultural property taxation,) or (c) enumerate the variety of impacts (e.g. trespass, farm fragmentation). The popular image of an agriculture in decline around cities is questioned and reasons for this stereotype suggested. These include a number of interpretative biases (e.g. that agriculture in peri-urban areas is responding to urban growth forces which are essentially negative for agriculture) and aspects of the methodology applied (e.g. use of secondary data sources, the scale of analysis and dangers of 'sampling at the margin', and emphasis on the land resource rather than the entrepreneur). It is suggested that what is required is more problem-specific data collected at the farm level. An interpretative framework which attempts to relate the various forces at work is presented in a threefold classification of the farming landscape. The second important research area, the societal response to urban impacts, concerns the need for, and the nature and role of, public intervention in the agriculture-urban development interaction process. For society, a critical question is not so much the quality of land being lost but the effect of this loss on the nation's food supply potential given particular political-economic goals. The importance of agricultural land must also be defined in relation to the value society attaches to other possible uses, and this raises the question of the appropriate criteria to evaluate policies. In conclusion, it is recommended that future research should give greater attention to investigation at the individual farm business scale to improve knowledge of the relative importance of different kinds of urban impacts. Secondly, more comparative research is required within a framework which acknowledges the interdependence of the factors operating in the peri-urban agricultural environment.

According to Marx there are two kinds of power, the power of property on the one hand and political power on the other. In Britain the importance of land as a factor of production in agriculture and the significance of agriculture in the countryside means that the question of land ownership and the agricultural land market is of fundamental importance to an understanding of rural economy and society. As well as being a factor of production, land is also capital. Thus in addition to private owners there are also several types of institutional landowners including traditional institutions (e.g. the Church and the Crown), public and semi-public bodies (e.g. nationalised industries), and financial institutions consisting of insurance companies, public and private pension funds and property unit trusts. Financial institutions have had a major impact on the UK agricultural land market over the last decade and this involvement has led to questions about their influence on the long term structure of agriculture and to concern over the growth of agribusinesses and the decline of traditional tenant farms. In Chapter 8 Peter Byrne explains the structure and operation of the agricultural land market and analyses the extent to which the effects of the land ownership pattern can be observed and interpreted, despite the sparsity of publicly-available statistical information. The distinction between land occupation and ownership and the complexity of the land tenure system with its various sub-groupings and arrangements is examined. Owner-occupied land, for example, may be farmed in hand with a manager, through a partnership, via a company, or farmed by the owner and his family directly. The reasons for owning farmland are then examined and the major types of landowners discussed. Information on agricultural land prices is more readily available than ownership data and the nature and problems of the three main sources are considered, prior to an examination of trends in land prices over the last decade. The relationships between market prices and land quality, tenure and size of holding are then examined. Predicting future trends in the agricultural property market and in the long-term structure of agriculture is problematic in view of possible changes in EEC and UK government policies, further technological developments in agriculture and the lack of a cadastral survey. Current trends suggest a stable proportion of tenant farms being reached by the early twenty-first century with the highest-quality and largest holdings being in the tenanted sector but owned by a smaller number of mainly institutional owners. It is suggested that unless appropriate legislation is enacted to stimulate private landowners to offer land for tenancy this sector could share the experience of the now largely residual private-rented housing market.

Traditionally, agricultural geographers have concentrated on the factors of production. Comparatively little attention has been paid to the processes of marketing and distribution despite the fact that the structure of the agricultural marketing system exercises a significant influence on the social and economic character of rural areas. In Chapter 9 Bill Smith stresses the incomplete nature of existing spatial theories of agricultural marketing. He emphasises the need for a dynamic analysis of the agricultural-food system which recognises the links between farmers, processors and retailers. The structure of the agricultural marketing system is then examined in detail and several models of the farmer-consumer system presented. A major consequence of the penetration of agriculture by corporate capitalism is that power is often located at some point between the farmer and consumer. The companies' wish to minimise market risks and improve corporate profits and stability often leads to oligopolistic control. The farmers' response to such a situation may include increased specialisation, larger farms, application of technology and replacement of labour by capital-intensive methods. Further research into the nature of the links between farm and market and the role of the price mechanism in the relationship is required. Vertical integration increases the flow of information and control between different levels of the agricultural-food system. Similar linkages are fostered by various forms of contract agreement between farmers and distributors under which the quality, volume, type and even timing of output may be determined by off-farm decision-makers. The nature of contracts is discussed before attention turns to the general question of market power. The positions of both small and large firms in the market are considered. It is observed that in some cases the relationship can be mutually beneficial as, for example, where small firms produce specialised goods thus ensuring their market viability and at the same time relieving consumer pressure on larger firms. Generally, however, the relationship between large and small firms is one of unequal competition. Attempts to provide small producers with some of the marketing advantages of large corporations have led to government-sponsored intervention via co-operatives and marketing boards, and the operation and effectiveness of these strategies is discussed. In conclusion, it is suggested that despite the complexity of market relationships and the absence of a conceptual framework the wide-ranging impact of market structure on the modern agricultural landscape emphasises that the analysis of market systems is an important emerging research issue in agricultural geography.

Famine and undernutrition are endemic to many parts of the world and in Chapter 10 David Grigg presents an overview of the world food problem. Precise definition of the scale and severity of the problem is hampered by data deficiencies and measurement problems but estimates based on the concept of food balance sheets suggest that one in five inhabitants of the Developing World suffers from an inadequate diet. Spatial analysis of levels of undernutrition reveals a group of countries, (whose combined populations account for one-third of the Developing World), in which the available food supply is insufficient to meet demand even if resources were distributed effectively, i.e. according to need rather than income. Although globally the growth in food output has exceeded that of population, for many individual countries food output *per capita* has declined. In countries, such as tropical Africa, where this trend is combined with the basic insufficiency of national food supply the outcome is often catastrophic. Population growth is not the sole cause of undernutrition, however, another fundamental factor being the poverty of the mass of the Third World population, and the relationship between income distribution and undernutrition is examined. The current situation in much of the Developing World is compared with the nutritional development of Europe. In the latter region uneven income distribution meant that although there was little evidence of undernutrition after the middle of the nineteenth century malnutrition remained in evidence until the Second World War. Two facets of the nutritional problem in the Third World can thus be identified. First, in countries where national food supply is sufficient to meet demand the incidence of undernutrition reflects the polarised distribution of national wealth. Resolution of this basic structural problem requires a fundamental re-ordering of socio-political organisation and, in the case of many former colonies, a re-orientation of their role in the world economic system. Secondly, in countries where total food supplies are less than national requirements the elimination of hunger requires both more equitable income distribution and increased output. Particular attention is then focused on means of improving food output *per capita*. These include expanding the area of cropland (e.g. through reclamation as in Latin America, reduced fallow as in Africa, or multiple cropping as in Asia), and improving yields (e.g. by increased labour inputs, irrigation or application of modern technology). It is concluded that while the proportion of the world's population suffering from undernutrition has decreased, in the Developing World the food problem remains formidable. Long-term structural remedies based on a redistribution of national wealth must be accompanied by more immediate measures to counter the low productivity of agriculture in order to balance existing food deficits and to absorb the continuing rapid growth in population.

1 THEORY AND METHODOLOGY IN AGRICULTURAL GEOGRAPHY

B.W. Ilbery

The basic objective of this introductory chapter is to emphasise the dynamic nature of agriculture and reflect upon the different approaches developed by geographers, especially since 1950, to aid the explanation of spatial variations in agricultural activities. Such is the size of the task, that only a generalised and broad overview can be presented. However, many of the issues raised will be clarified and further developed in subsequent chapters. It will hopefully be shown how the development of theory and methodology in agricultural geography is essentially a reflection of the changing philosophy of, and progress in, human geography in the post-war period.

The Nature of Agricultural Geography

Agricultural geography seeks to describe and explain the distribution of farming activities over the earth's surface. Therefore, it comprises two parts: the first has location and context as central themes and is concerned with recognising and analysing spatial variations in agricultural and farming practices throughout the world (Coppock, 1968); the second attempts to explain the great diversity of agriculture. This latter task is a complex one and, in the absence of data on social and economic aspects of farming, explanation was often sought in terms of physical and historical factors (Coppock, 1964). In reality, a proper insight into the distribution of farming types can be obtained only by examining the nature of the relationships between a large number of influencing factors. These relationships are of many different kinds, which cannot be incorporated into a single system of laws (Morgan and Munton, 1971), and consequently it has proved difficult to develop a truly realistic model of agricultural land-use. In common with most geographical studies, agricultural geography has a scale problem. With an increase in scale, from micro to macro, the roles of influencing factors change and physical controls become more important than management and personal factors in agricultural land-use patterns. As a generalisation, agricultural geographers have reduced their scale of analysis over time, with the result

13

that economic and social factors have been increasingly stressed at the expense of broad-scale environmental influences. This has been reflected in a movement of interest away from the delimitation of large-scale and physically determined agricultural regions (Whittlesey, 1936) and towards the farm as the basic decision-making unit in agriculture (Haines, 1982).

If one reviews the mounting literature on agricultural geography, two major approaches to the subject matter can be detected (Ilbery, 1979):

1. *An empirical (inductive) approach,* which attempts to describe what actually exists in the agricultural landscape. Explanation of the patterns is sought by inductive methods and generalisations are made on the basis of results from numerous studies.
2. *A normative (deductive) approach,* which is more concerned with what the agricultural landscape should be like, given a certain set of assumptions. This approach leads to the derivation and testing of hypotheses and, theoretically, to the development of an ideal model of agricultural location.

These two approaches have never really merged, reflecting both the complexities of the decision-making process in agriculture and the different times at which each has been popular within geography. It is essentially from the normative approach that models of agricultural location have emerged and once again model makers have operated along one of two lines, with the latter developing out of dissatisfaction with the former:—

1. *Optimiser models,* which are usually concerned with the notion of profit maximisation. Optimal land-use patterns rest on the assumptions of farmer rationality, complete knowledge and an equal ability to use this knowledge. Such requirements are unobtainable in reality and the approach has been criticised as unrealistic. Farmers cannot make perfect economic decisions, except by chance, and instead react to perceived conditions within an environment of uncertainty. Consequently, 'satisfaction' has been put forward as an alternative, leading to
2. *Satisficer models,* which are more realistic and take farmers' motivations, aspirations and attitudes into account. These models include such items as a farmer's desire for leisure, a satisfactory income and social considerations, at the expense of profit maximisation. This approach developed out of two classic geographical studies, by Wolpert (1964) in an analysis of Swedish farming and Harvey (1966) in an early review of theoretical developments in agricultural

geography, and led to a new behavioural element in the methodology of the subject.

As with normative and empirical approaches, there is a noticeable gap between these two groups of models and it would appear that even satisficer models are failing to explain the observed world adequately. Despite these different approaches and the many methods available to the geographer, theoretical developments in agricultural geography have been slow. Indeed, it could be suggested that little real theoretical progress has been made since the pioneering work of von Thünen (1826). However, different modes of explanation exist and following Tarrant (1974) these can be categorised into three main theoretical approaches, which demonstrate how the focus of interest has shifted through time in the post-war period.

1. *Environmental,* which assumes that the physical environment acts in a deterministic manner and controls agricultural decision-making.
2. *Economic,* which assumes that the economic factors of market, production and transport costs operate on a group of homogeneous producers, who in turn react to them in a rational manner.
3. *Social-personal,* which assumes that there are further sets of influences which affect agricultural decision-making, including farmers' aims, values, motives and attitudes towards risk.

With physical influences being increasingly modified by man, through such items as fertilizers, irrigation and early ripening varieties of crops, it is not surprising that economic and social factors have been emphasised in the spatial structure of agriculture. Indeed, the economic environment of farming is dynamic with a number of emerging trends, ranging from the modernisation and industrialisation of agriculture, through an increase in large scale capital-intensive farming, specialisation of production and greater integration with food processing industries, to the growth in part-time farming or mulitple-job holding (Bowler, 1984; Grigg, 1983). These trends, and especially the way in which they are perceived by farmers, will have far-reaching effects on the agricultural landscape and must be incorporated into methodological and theoretical developments within agricultural geography.

Attention in this chapter will now be focused upon each of the three theoretical perspectives, with the methods adopted in each approach being particularly emphasised.

The Physical Environment

Recognising and delimiting agricultural regions was once a central task of agricultural geographers. This occurred on numerous scales, from the world (Whittlesey, 1936) down to the individual county in England (Tavener, 1952). Whichever scale of study was adopted, the resultant regions were often formed on the basis of such physical criteria as soils and climate. Implicit in this work was that differences in the physical environment determined spatial variations in agricultural activities. Such geographical determinism was popular for many years, until first the USA and then Britain and Europe reacted against it.

As Grigg (1982a) rightly pointed out, it has been surprisingly difficult to establish an environmental theory of agricultural location, even though the importance of environmental variations influencing crop and livestock distributions is readily apparent. The same author attributes this to the large number and complexity of the influences that determine the spatial arrangement of crop and livestock types, many of which are physical in nature. Therefore, whilst the physical environment cannot be ignored, it is the interaction of physical and human factors that determines patterns of agricultural land-use. This interaction is in turn made more complex by two further sets of factors (Tarrant, 1974): first, the personal characteristics of the farmer, including his knowledge of new agricultural ideas and innovations and his attitudes toward risk avoidance; and secondly, the dynamic nature of agriculture, which may lead to an imbalance between physical and economic environments. The complexity of the situation is enhanced because farmers will perceive and react in different ways to changes in mechanisation, methods and crop varieties.

It needs to be emphasised that there are very few situations, if any, where physical factors are either all-important or of no account in the distribution of agricultural practices. Two contributory factors have to be considered: scale and type of crop. As previously mentioned, it is possible to relate regional but not local variations in agriculture to broad environmental contrasts. Similarly, certain crops require specific physical and biological conditions, whilst others can be grown in a range of physical environments. This is one of various factors ignored by Ricardo (1817), who assumed that variations in economic rent, defined as the return which can be realised from a plot of land over and above that which could be realised from a plot of the same size at the margin of cultivation, were caused by differences in soil fertility. Consequently, margins of cultivation were determined solely by physical factors, a point that could not be verified by empirical case studies.

The different physical requirements of different crops can be used to demonstrate two important and interrelated concepts in relation to physical factors: the ecological optimum and the margin of cultivation. Introduced by Klages (1942), the ecological optimum argues that for any crop there are minimum requirements of moisture and temperature without which the crop will not grow, and also maximum conditions beyond which growth ceases (Grigg, 1982a). These requirements vary from plant to plant and provide a spatial limit (margin of cultivation) beyond which a crop cannot be grown. Maize in the mid-west region of the USA is often used to demonstrate this concept. The optimum conditions for maize decline in all directions from a central point: to the north and south it becomes too cold and too hot respectively, whilst to the west and east it becomes too wet and too dry. Farmers will rarely cultivate maize near these physical margins, unless changing economic considerations, such as a fall in production costs or an increase in price, cause the economic margin to extend towards the absolute limit.

These points can be further developed with reference to case-studies from Scotland and Finland. Parry (1976) argued that changing economic conditions could lead to an increase in cultivation in marginal areas, usually in former cultivated areas now abandoned. Taking the Lammermuir Hills in south-east Scotland as his study area, Parry showed that over 11,400 hectares (ha.) of existing moorland had been improved at one time, 4,890 ha. before 1860 and 6,500 ha. between 1860 and 1970. In an earlier study (Parry, 1975), the distribution of this abandoned land was related to secular climatic change. Using three indicators of climatic change — exposure, summer wetness and summer warmth — Parry demonstrated how the climatic limit to cultivation may have fallen 140 metres over 300 years. By comparing the upper and lower climatic limits of marginal land with the distribution of abandoned land before 1800 (Figure 1.1) a strong correlation was obtained, indicative of an indirect causal relationship. However, Parry was careful to point out that land abandonment in the marginal areas of south-east Scotland was due only in part to the deterioration in climate. A proper examination of abandoned land in the area would also have to consider soil exhaustion and such human influences as the decline in the monastic farming system and fluctuations in the demand for agricultural products.

The idea of physical margins to cultivation being partly controlled by economic factors was clearly demonstrated by Varjo (1979) in Finland, a country crossed by the absolute limits of numerous crops. Varjo plotted the changing cultivation limits of barley, oats, rye and spring wheat between 1930 and 1969 and hypothesised that the progressive southward

Figure 1.1: Land Abandonment in South-east Scotland Before 1800

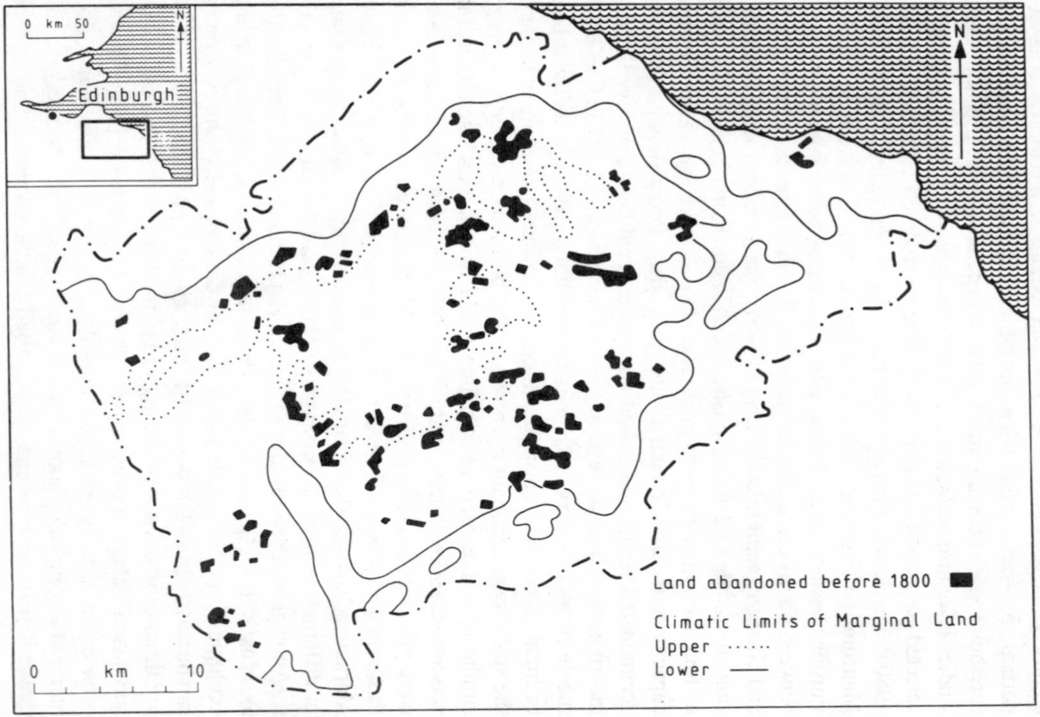

Land abandoned before 1800 ■

Climatic Limits of Marginal Land
Upper ·········
Lower ———

Source: Parry, 1975, pp. 4 and 6.

movements were responses to deteriorating climatic conditions. However, the climate had worsened only in the 1940s and the southward movement of the limits of production was due to the sharp decline in profitability of cereals.

Although relatively neglected over the last twenty years, there has been a revival of interest in the effect of environmental influences on the spatial structure of agriculture. To a certain extent this is a reflection of economic recession and the price-cost squeeze in agriculture. As a consequence, yields and intensity of production have become very important, encouraging farmers to seek the 'best' physically endowed areas for their enterprises (Winsberg, 1980; Bowler, 1981). Many recent studies have emphasised the effects of physical factors on the yield of selected crops. These relate to such diverse areas as California (Granger, 1980a, b), South Africa (Gillooly, 1978; Gillooly and Dyer, 1979), Mauritania (Vermeer, 1981), North America (Michaels, 1982) and Europe (Dennet *et al.*, 1980). Briggs's (1981) study of the relationships between the yields of spring barley in England and Wales and edaphic and climatic conditions is typical of their work. Using multiple regression, it was demonstrated how the yields of spring barley are related to four factors: sand content, drainage conditions and available water capacity of the soil, potential summer soil moisture deficit and annual accumulated temperature. Briggs argued that these variables should be incorporated in soil survey and land-use capability classifications, so that the discrepancy between actual and potential yields of cereal crops can be highlighted. With such information being made available to farmers, the spatial pattern of spring barley could be fundamentally changed. A similar spatial relocation, of potato production in the United Kingdom, was advocated by Ingersent (1979), who argued that climatic conditions were primarily responsible for the instability of potato yields. However, this is an impractical suggestion as production is controlled by the Potato Marketing Board, which is unable to discriminate between its members in different parts of the country.

Another recent approach is to regard agricultural and farm systems as ecosystems (Bayliss-Smith, 1982; Simmons, 1979) and modern ecological models, which emphasise the flow of energy and nutrients, can be adopted to analyse agricultural systems. The idea of energy flowing through an agricultural ecosystem is not a new one, but the costing and quantification of energy used in agricultural production is. This approach gained impetus as a result of the energy crisis, when the sudden increase in oil prices highlighted the dependence of advanced farming methods upon energy from fossil fuels (Bayliss-Smith, 1982). Wood

(1981) suggested two themes concerning energy and agriculture that are of particular geographical interest: first, a description, discussion and explanation of spatial variations in levels of energy efficiency; and secondly a focus of attention on the range of responses that have already occurred or are expected with worsening energy price/supply conditions. Responses to the energy crisis could have far-reaching consequences for patterns of agricultural production and whilst agriculture is not a major consumer of energy, it is inefficient in its use, as demonstrated by Noble (1980) for British horticulture and glasshouse production in particular and by Buttel and Larson (1979) for corporately-owned farms in the USA.

Actual changes in behaviour and land-use patterns which have resulted from energy price rises still await attention from geographers, but Bayliss-Smith (1982) has warned that a strictly ecological perspective does not provide a very coherent framework for agricultural geography. In line with many other researchers, Bayliss-Smith advocates a more integrated approach, which relates ecological pressures to farmers' responses, goals and values and the structure of social organisations. Indeed, Merrill (1976) has proposed a more radical approach to studying the agricultural landscape, whereby researchers should assume that fossil fuels are exhaustible and that an important role of agriculture is to sustain farmlands with ecologically-wise methods. There is a growing interest in landscape conservation and consideration is being given to the detrimental effects of increased specialisation and intensification in agriculture.

One particular area of interest has been the loss of agricultural land to urban development. The reason for mentioning this in the physical section is that much of the land being lost is of good physical quality (Edwards and Wibberley, 1971). In England, urbanisation is concentrated in the south and east where the better land is also found (Coleman, 1978). Similarly, Gregor (1963) and Platt (1977) noted a tendency towards the urbanisation of the better soils in California; the latter estimated that 21 per cent of California's prime agricultural land has been urbanised. Reasons given for this trend included the gentle terrain and the fact that urban centres originated as service centres for the agricultural community. In the Niagara fruit belt, it has also been established that urbanisation has occurred on land where the ratio of 'tender fruit' soils to 'tender' climate is most favourable (Krueger, 1978).

The Economic Environment

As agriculture is primarily an economic activity governed by the laws of supply and demand, there is a need for a systematic understanding of the importance of economic considerations. Economic theory has assumed that farmers are rational profit maximisers who respond automatically to changes in prices. These principles were applied by Ricardo (1817) and von Thünen (1826), who both used the concept of economic rent to determine patterns of agricultural production. However, it was von Thünen who devised the first economic model of agricultural location, arguing that 'distance to the market was the prime determinant of which crops and livestock were grown and with what intensity' (Grigg, 1984, p. 17).

Therefore, unlike Ricardo's ideas, which were based on production advantages, von Thünen's model rested on transfer advantages, where economic rent was controlled by distance from the market and transport costs. The actual model, which has been fully described in standard texts (Morgan and Munton, 1971; Symons, 1978; Tarrant, 1974), was based on empirical evidence of agriculture in his own locality and intended as a method of analysis rather than a theory of location (Chisholm, 1979). Economic rent was shown to decline with increasing distance from the market and this provided the necessary framework for both intensity and crop aspects of the model (Norton, 1979). The former stated that the intensity of production of a particular crop would be inversely proportional to distance from the market, whilst the latter used different economic rent curves to demonstrate the tendency for agricultural land uses to become concentrated in concentric rings around a central market.

Despite the many limiting assumptions of his work and especially since its translation into English in 1966, von Thünen's ideas and the effects of distance on land-use patterns in particular continue to interest agricultural geographers. The zonation of agricultural activities, in relation to distance from a farm or central market, has been reported on numerous scales (Chisholm, 1979). These range from De Lisle's (1982) study of intra-farm variations in cropping patterns among the Mennonite cultural group in Manitoba and Blaikie's (1971) and Richardson's (1974) investigations of agricultural practices around north Indian and Guyanan villages, to the work of Golledge (1960), Horvarth (1969) and Griffin (1973) on land-use around the cities of Sydney, Addis Ababa and Montevideo, and surveys by Ewald (1976) and Van Valkenberg and Held (1952) in colonial Mexico and north-western Europe. Two further applications of von Thünen's ideas are worthy of mention. The first is the development of the model in a dynamic context, as demonstrated in Peet's (1969) classic paper on imports into Great Britain in the nineteenth century, which came from an ever-expanding but logical Thünen system, and Day and Tinney's (1969) application using linear programming. The second is macro-scale zoning of land-use, which assumes an extensive urban area and a continent-wide hinterland and was reported for western Europe by Valkenberg and Held (1952) and Belding (1981), and for the USA by Muller (1973) and Jones (1976). However, evidence of large-scale land-use zoning is not always conclusive and Kellerman (1977) could find no such pattern in the USA, despite the findings of Muller and Jones.

Although many studies have been produced on the application and modifications of von Thünen's model, the relative importance of distance and transport costs on agricultural location has declined. Transport costs are offset by tapering effects and physical distance is less important than time, cost or perceived distance. Similarly, agriculture has witnessed numerous technological developments which, along with the major improvements in modes of transport, have relegated distance and transport costs to a lower ranking in the list of decision-making factors affecting farmers. Indeed, Sinclair (1967) argued that rapid urban development, rather than transport costs to the market, would affect the intensity of agricultural production in the vicinity of cities. He demonstrated how the level of investment in agriculture would be reduced in peri-urban areas, as land speculators pushed prices beyond the reach of ordinary farmers. Much land would lie idle which, together with a growth in part-time and hobby farming, could lead to more extensive agriculture in the rural-urban fringe. Therefore, unlike von Thünen, Sinclair claimed that the intensity of agricultural production would increase with distance from the market.

The Sinclair model, or reverse approach, has similarly attracted much attention, gaining considerable support from numerous authors (see Berry, 1979; Mattingley, 1972; Bryant, 1974). However, many studies have emphasised the complex interplay of factors at work in the rural-urban fringe, creating varied patterns in different areas. For example, in a study of Hong Kong, Sit (1979) found evidence to support aspects of both direct (von Thünen) and reverse (Sinclair) approaches to fringe farming. Similarly, the formal and informal controls of institutional restrictions have helped to produce a heterogeneous pattern of agricultural land-use in Auckland's rural-urban fringe (Moran, 1979). Only a weak inverse relationship could be discerned between intensity of production and distance from the city and Moran pointed to a number of factors which helped to account for the complex pattern. These ranged from natural physical advantages and the subdivision of land for orcharding, to the distribution of Yugoslav and Chinese ethnic groups for market gardening, the allocation of milk supply quotas, special rate assessments and a strong farm lobby. Milk quotas are allocated to give preference to farmers close to the urban area and special legislation has been passed to assess rates on a different basis for land that is farmed, compared with other land on the urban periphery. These policies have no doubt been partly governed by the strong farm lobby, which has helped to influence the direction of urban growth and maintain particular parcels of land in agricultural production.

Whilst distance may have declined in importance, the market remains one of the most potent factors in agricultural production. However, owing to the large number of independent producers, the individual farmer has no control over the price he receives for his goods. Faced with such a marketing problem, farmers have three main options open to them: first, they can call for direct government intervention; secondly, they can group together and form their own co-operative marketing system; and thirdly, they can circumvent the wholesale-retail channels by negotiating contracts with food processors or marketing their produce direct to the public. Although there has been some geographical interest in forms of direct marketing (Bowler, 1982; Linstrom, 1978), the spatial consequences of these three courses of action have been under-researched.

Government intervention in agriculture has taken various forms, ranging from price and production controls to marketing boards, structural reform, and grant aid and income supplement. Each has had spatial expressions, which have been detailed by Bowler (1979). It has increasingly been recognised that a policy of protection, as practised for many years under the Common Agricultural Policy of the EEC, is not going to solve the problem of low farm incomes. As farm structure is a major determinant of income, structural reform has been seen as the only long-term solution to poor incomes. Many schemes exist in Europe, Australia and North America, including land consolidation, farm enlargement and land reform (Bowler, 1983; King, 1977; King and Burton, 1983). Structural reforms have been practised in France, for example since 1941 and their spatial consequences have been examined by various authors (Baker, 1961; Clout, 1968, 1972, 1975; Naylor, 1982).

Much has been written on the economics and theoretical benefits of agricultural co-operation (Le Vay, 1983). However, reasons for the distinct spatial variations in its success have not been analysed in detail, although local and national traditions play a part, just as co-operation lends itself to some crops better than others, especially those of a horticultural nature. Co-operatives have been far more successful in the USA, on the European continent and in some developing countries than they have in the United Kingdom. This led Hewlett (1967) to conclude that interest in co-operative movements has been in inverse proportion to the level of contemporary agricultural prosperity. Reasons for the relative lack of agriculture co-operation in the United Kingdom include the general prosperity and heterogeneous nature of British post-war farming, the independent and conservative nature of the farmers, the insignificant export trade in agricultural produce and the lack of government support for co-operatives until 1967. Within Britain, agricultural co-operation has been most successful in areas of large, mainly arable farms and hence the eastern counties of England and Scotland (Bowler, 1972).

In the post-war period, the agricultural economics of many western countries have undergone a process of profound change, commonly referred to as the third agricultural revolution. This has been characterised by 'the progressive extension of technological, organisational and economic rationality into the arena of farm operations, linking them even more closely to the other sectors of the economy both materially and in ethos' (Wallace, 1985, p. 6). Whilst no model of the modernisation and industrialisation of agriculture has been forthcoming, technological change and the trend to fewer, larger and more capital-intensive farms have received certain attention from agricultural geographers (Gregor, 1979, 1982). Three aspects in particular have been examined: first, economies of scale and increasing farm size; secondly, enterprise and regional specialisation; and thirdly, the relationship between farmers and food processing industries.

Although external economies have been important in creating areas of localised agricultural production, as in the hop industry in Kent (Harvey, 1963), horticultural production in the Vale of Evesham (Buchanan, 1948) and poultry production in the East Midlands (White and Watts, 1977), it is internal economies of scale, from increasing farm size and specialisation, which have received most attention. Increasing farm size is a generally accepted trend in most advanced economies and, as there is a direct relationship between farm size and enterprise type, spatial variations in the distribution of farm sizes are reflected in patterns of agricultural land-use. Farm enlargement is caused mainly by the process of amalgamation, which has been shown to vary spatially according to certain factors (Clark, 1979; Kampp, 1979; Todd, 1979). In turn, increasing farm size has heightened the problem of farm fragmentation (King and Burton, 1982). The relationship between fragmentation and farm size has been demonstrated by Smith (1975) and Carlyle (1983) in North America, Hill and Smith (1977) in Australia, and Edwards (1978) and Ilbery (1984a) in England.

Agriculture has become concentrated in the hands of fewer and larger producers and both concentration and specialisation of production have had marked locational effects on patterns of agricultural land-use. In the USA and UK, the majority of farming types have become more regionally concentrated (Bowler, 1975a, 1981; Winsberg, 1980), with 'mixed farming continuing to give way to more specialised types' (Britton, 1977, p. 203). The factors at work in this process are complex, but the differential movement in and out of production of certain products is a key element (Bowler, 1981).

One characteristic of the decline in number of farm enterprises is the virtual elimination of pig and poultry enterprises, which have become increasingly specialised into industrial-type units. Indeed, large, often transnational, corporations have become involved in one or more stages of the agri-food production system, indicating a move from agriculture to agribusiness. The latter can be defined as agriculture organised around scientific, rational and industrial business principles. At the heart of the concept is the relationship between the agricultural production industry and two related sets of industries: upstream, the agricultural supply industries and, downstream, the food processing industries. Whilst geographers have yet to become fully involved in the spatial aspects of agribusiness, it is clear that vertical integration in agriculture is much further developed in the USA than in Britain. For example, Smith (1980) has plotted the distribution of America's richest farms and ranches, three-fifths of which are fully integrated agribusinesses. The distribution shows areas of economic power for different agricultural types, with California and Florida exhibiting the most striking concentrations of wealthy agricultural enterprises in the late-1970s (Figure 1.2). In Britain, integration is characterised more by agribusiness companies seeking out highly market-oriented farmers with whom to place contracts. Therefore, contract farming favours the larger farms of eastern England and is particularly important for certain frozen vegetables, notably peas (Dalton, 1971; Hart, 1978).

The Behavioural Environment

The failure of traditional economic models to provide realistic explanations of agricultural land-use patterns led, in the 1970s, to a greater consideration of behavioural factors in the spatial structure of agriculture (Hart, 1980; Ilbery, 1978). Based on the assumption that there are further sets of influences which affect agricultural decision-making, including farmers' values, aims, motives and attitudes, the behavioural approach recognises that farmers may not always perceive the environment as it

Figure 1.2: America's Richest Farms and Ranches — late 1970s

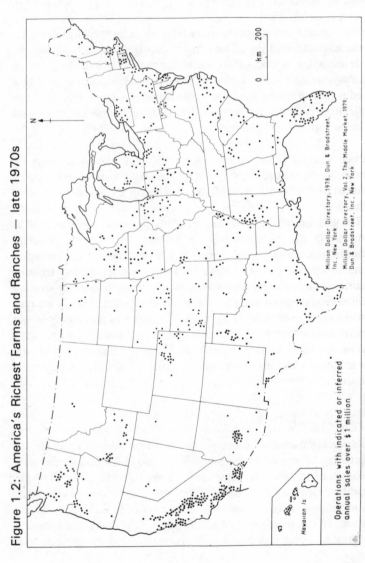

Operations with indicated or inferred
annual sales over $1 million

Million Dollar Directory, 1978, Dun & Bradstreet,
Inc. New York

Million Dollar Directory, Vol.2, The Middle Market, 1979,
Dun & Bradstreet, Inc. New York

Source: Smith, 1980, p. 53.

is. Indeed, the fundamental unit of study in agricultural geography is the farm and the farmer, a fact that was often overlooked in previous physical and economic approaches to the subject.

Therefore, the objective of the behavioural approach is clear — to reject the notion of economic man and replace it with a model that is closer to reality. The situation was admirably summarised by Harvey (1966, p. 373):

> if we recognise the all-important fact that geographical patterns are the result of human decisions, then it follows that any theoretical model developed to explain agricultural location patterns must take account of psychological and sociological realities, and this can only be achieved if the normative theories of agricultural location are made more flexible and blended with the insights provided by models of behaviour.

Variations in economic behaviour cannot be explained in terms of the availability of resources (Wolpert, 1964) and a greater appreciation is required of the importance of social conditions and human motives in farming.

The decision-making behaviour of farmers can be viewed as a reflection of a wide range of values (Gasson, 1973), from family security and a satisfactory income to being creative and independent and belonging to the farming community. However, many factors affecting decision-making are unpredictable, increasing the degree of risk and uncertainty involved and demonstrating the large chance element in the determination of land-use patterns (Hart, 1980). The causes of uncertainty are extensive, ranging from technological change and government policy to disease, the climate and such personal factors as a farmer's health, age and ability to work. All farmers make choices within this environment of uncertainty, preventing the attainment of an optimising goal. In addition, actual decisions made will vary because farmers have different goals, levels of knowledge and perceptions of and attitudes towards risk.

Game-theoretic models attempt to provide a normative solution to decision-making, in the light of farmers' incomplete knowledge and uncertainty. They introduce probabilistic formulations into decision-making, associated with conditions of uncertainty caused by such variables as weather patterns and market prices. Therefore, game theory is concerned with the rational choice of strategies in face of competition from an opponent, usually the environment. A number of solutions to the developed 'pay-off' matrix, which shows the outcome of each move by the farmer against each possible move by the environment, are possible (Agrawal and Heady, 1968). The choice of solution depends very much upon the type of farmer concerned and his attitude towards risk avoidance. In an early application of game theory, Gould (1963) determined the choice of strategies (crops) which would help to win the basic struggle for survival in the barren Middle Zone of Ghana. A more recent and novel application of game theory by Cromley (1982) attempted to develop the principles in relation to von Thünen's land-use model. Patterns of land-use were deemed to vary according to the goals and values of farmers and Cromley tried to examine deviations from the idealised system which resulted from uncertain weather conditions. However, game theory can become mathematically complex and farmers are unlikely to incorporate elements of it in their decision behaviour knowingly. Consequently, a number of criticisms have been levelled against game-theoretic models (Ilbery, 1985; Tarrant, 1974).

Another important set of decision-making models concerns the diffusion or spread of innovations, their adoption or non-adoption and resultant effects on patterns of land-use. By the late-1960s, innovation diffusion research had become prominent in the establishment of the behavioural approach in geography. Traditional approaches to diffusion studies focused upon the processes by which adoption occurs, or the demand aspect of diffusion. These have been synthesised by Jones (1967) and Rogers and Shoemaker (1971) and represented in geography by the Monte Carlo simulation model of Hagerstrand (1967). Hagerstrand conceptualised the adoption of innovations as the outcome of a learning or communications process, in which interpersonal information flows were very important. His model was able to produce a series of maps, depicting the spatial distribution of the adoption process over time, and many subsequent studies were based on this idea, including Bowden's (1965) work on the adoption of pump irrigation in Colorado, Misra's (1969) survey of agricultural innovations in Mysore, India, and Johansen's (1971) investigation of strip cropping in south-western Wisconsin. An important empirical regularity in these and many other similar studies was the logistic curve of adopters, which could be used to divide adopters into different categories. In turn, the major differences in the economic, social, locational and demographic characteristics of each group could be revealed.

The reliance of this 'adoption perspective' upon personal information flows and characteristics of the individual decision-maker has been criticised by the more recent literature on innovation diffusion. Attention has increasingly been concentrated upon the role of supply factors and constraints in the diffusion process. This is because individual choice behaviour is constrained by government and private institutions (Brown, 1981). Therefore, the first stage in the diffusion process is the establishment of diffusion agencies which develop and implement strategies to promote adoption in their market areas. This kind of 'market and infrastructure perspective' was developed by Brown (1975) and has been applied in many case-studies, including a number from Third World countries (see Brown, 1980; Brown *et al.*, 1977; Brown and Letnek, 1973; Garst, 1974; Havens and Flinn, 1975; Yapa and Mayfield, 1978). In an attempt to focus attention more on human cognitive processes, Brown (1980) compared attitudes toward adoption with a set of socio-economic variables and found, in a study of five agricultural innovations in Ohio, that attitudinal variables were superior to social-category variables in their ability to discriminate between adopters and non-adopters.

The behavioural approach has indeed attracted considerable interest in attitudes and agricultural geographers have developed or made use of different techniques to elicit such information from farmers. An early study by Butler (1960) on the attitudes and motives of farmers used the concept of a 'modal' farm to distinguish farm units that deviated from the 'norm' for the area. Explanations for any deviations were sought in terms of physical, social and economic factors. Similar approaches, but using trend surface residuals as deviations from the norm, were adopted in studies of the trend to enterprise specialisation by Bowler (1975b) and Ilbery (1984b).

A different approach was adopted by Gasson (1973) who developed a methodological framework for analysing farmers' goals and values. Gasson was concerned with farmers' motivations, in an attempt to discover what farmers actually wanted from their occupation. Emphasis was placed more on why, than the way, decisions are made, on the basis that a better theory of farmers' behaviour would incorporate behaviour as conditioned by customs, habits, perceptions, beliefs and values. A list of 20 values, categorised into four groups, was provided that might apply in a broad range of farming situations. These were tested in East Anglia and amongst hop farmers in the West Midlands (Ilbery, 1983). In both cases (Table 1.1), intrinsic values were emphasised above expressive and instrumental values, with social values having the lowest priority of all. Therefore, farmers would appear to place more importance upon doing the work they like and being independent than on the income aspects of farming.

Other techniques used to elicit farmers' attitudes include repertory grid procedures and point score analysis. The former have been used

Table 1.1: Goals and Values of Farmers in East Anglia and Hop Farmers in the West Midlands

Values	Rank Order	
	East Anglia (Gasson, 1973)	West Midlands (Ilbery, 1983)
Intrinsic		
Doing the work you like	1	1
Independence	2	2
Healthy, outdoor life	5	7
Purposeful activity	—	14
Control in variety of situations	—	8
Expressive		
Meeting a challenge	4	6
Being creative	8	10
Pride of ownership	—	15
Self-respect	11	5
Exercising special abilities	—	—
Instrumental		
Making maximum income	9	9
Arranging hours of work	14	14
Expanding the business	6	13
Safeguarding income	7	8
Making satisfactory income	3	3
Social		
Belonging to farming community	15	12
Working near family	10	11
Family tradition	13	15
Prestige of farmer	—	20
Respect of workers	12	4
Overall rank order		
Intrinsic	1	1
Expressive	2	2
Instrumental	3	3
Social	4	4

Source: Gasson, 1973, p.529 and Ilbery, 1983, p. 333.

to examine the perceived world of the farmer, especially in a Third World context (Floyd, 1976; Townsend, 1977), and the latter was developed to assess the relative importance attached by the farmers themselves to physical, economic and socio-personal factors in the decision-making process (Ilbery, 1977). These and many similar studies have confirmed that other motives are as important as maximum profit (Grigg, 1984). In particular, a major objective of farmers is to obtain a secure and stable farm business, which involves social considerations as well as economic and physical considerations (Ilbery, 1978). A second motive is the independence that farming as a livelihood gives, encouraging owners of many small farms to continue with their own system when they could earn more as labourers on larger and technologically more modern farms.

The rural-urban fringe has already been mentioned in previous sections of this chapter and for a long time studies of this dynamic area developed an essentially economic perspective. However, in line with the behavioural approach, it was increasingly recognised that land-use patterns in the fringe could not be fully appreciated without consideration of farmers' attitudes and responses to the threat of urban development; greater attention should be given to the motivation of landowners and occupiers than to searching for patterns of intensity (Moran, 1979). These aspects were examined by Bryant (1981), in a study of farmers' responses to urban development in the fringe of Paris, Blair (1980), in a survey of urban influences on Essex farming, and Layton (1981), in a comparison of attitudes between commercial and hobby farmers in the rural-urban fringe of London, Ontario.

It should be clear from this section that factors of a non-physical and non-economic nature may help to explain the distribution of farming types in the western world. However, the behavioural approach has similarly been criticised and the relative 'newness' of the perspective can only allay a portion of the criticism. Grigg (1982b, p. 243) has described the behavioural approach as 'a subject where there seems as yet to be more methodology than results' and Bowler (1984, p. 259) has stated that 'when applied at a local or regional level, the relative role of behaviour in relation to other factors is more difficult to discern'. It is, of course, difficult to observe human behaviour objectively, especially as farmers may be unaware of the bases of their decisions. Bunting and Guelke (1979) have voiced further criticism of the behavioural approach, in particular its failure to solve the explanation problem. They commented upon the slowness to develop realistic theories and the failure to relate attitudes to observed patterns of behaviour. A lack of coherence is apparent and there is over-concentration on studies of images and preferences and the way these relate to the socio-economic characteristics of individuals. However, in defence of the behavioural approach, many empirical studies have revealed genuine insights into the importance of behavioural influences in farmer decision-making, even if the universal application of their results has not been demonstrated. It is only from such observations that a coherent conceptual framework can be developed to explain agricultural land-use patterns and one in which behavioural influences cannot be ignored.

Conclusions

It has been the intention of this introductory chapter to provide a skeletal outline of some of the trends in the development of theory and methodology in agricultural geography. The complexity of the agricultural landscape has meant that theoretical progress has been rather slow. However, explanations that have been developed can be grouped into three broad categories, which represent a continuum of thought in post-war human geography.

Whilst the focus of interest has shifted through time, it would be wrong to give the impression that only one mode of thought was followed at any one time. For example, the growth in importance of the behavioural approach during the 1970s did not lead to the exclusion of studies adopting either physical or economic perspectives. Agricultural land-use patterns result from the interaction of all three sets of factors and a major problem has been that individual researchers have tended to concentrate on one approach, to the exclusion of other perspectives; more research projects of an integrated nature are urgently needed.

Whichever approach is adopted, agricultural geographers are being increasingly drawn towards a number of agricultural 'issues' which attract interdisciplinary attention. According to Bowler (1984), these can be divided into four broad groups: first, the modernisation and industrialisation of farming systems, especially structural change in agriculture and the increased scale of farming, but also developments in agribusiness; secondly, the loss of agricultural land in developed economies for urban development; thirdly, the growth in importance of institutional determinants, especially government intervention and changes in land tenure; and fourthly, the growth and importance of part-time farming in western countries. To this list can be added the revival of interest in the effects of environmental influences on crop yields, the role of energy in farm systems and the conservation of agricultural resources. This makes a fairly impressive catalogue of concerns for agricultural geographers, incorporating physical, economic and behavioural considerations, and ensures a dynamic future for agricultural geography and its links with such related subjects as rural sociology and agricultural and industrial economics.

References

Agrawal, R.C. and Heady, E.O. (1968) 'Application of game theory models in agriculture;

Journal of Agricultural Economics, 19, 207–18

Baker, A.R.H. (1961) 'Le remembrement rural en France', *Geography, 46,* 60–2

Bayliss-Smith, T.P. (1982) *The ecology of agricultural systems,* Cambridge University Press, Cambridge

Belding, R. (1981) 'A test of the von Thünen locational model of agricultural land use with accountancy data from the European Economic Community', *Transactions of the Institute of British Geographers, 6,* 176–87

Berry, D. (1979) 'Sensitivity of dairying to urbanisation: a study of North-East Illinois', *Professional Geographer, 31,* 170–6

Blaikie, P.M. (1971) 'Spatial organization of agriculture in some North Indian Villages', *Transactions of the Institute of British Geographers, 52,* 1–40

Blair, A.M. (1980) 'Urban influences on farming in Essex', *Geoforum, 11,* 371–84

Bowden, L.W. (1965) *Diffusion of the decision to irrigate,* Research Paper Series, Department of Geography, University of Chicago

Bowler, I.R. (1972) 'Co-operation: a note on governmental promotion of change in agriculture', *Area, 4,* 169–73

Bowler, I.R. (1975a) 'Regional variations in Scottish agricultural trends', *Scottish Geographical Magazine, 91,* 114–22

Bowler, I.R. (1975b) 'Factors affecting the trend to enterprise specialisation in agriculture: a case study in Wales', *Cambria, 2,* 100–11

Bowler, I.R. (1979) *Government and Agriculture: a Spatial Perspective,* Longman, London

Bowler, I.R. (1981) 'Regional specialisation in the agricultural industry', *Journal of Agricultural Economics, 32,* 43–54

Bowler, I.R. (1982) 'Direct marketing in agriculture: a British example', *Tijdschrift voor Economische en Sociale Geografie, 73,* 22–31

Bowler, I.R. (1983) 'Structural change in agriculture' in M. Pacione (ed.), *Progress in Rural Geography,* Croom Helm, London, pp. 46–73

Bowler, I.R. (1984) 'Agricultural geography', *Progress in Human Geography, 8,* 256–62

Briggs, D. (1981) 'Environmental influences on the yield of spring barley in England and Wales', *Geoforum, 12,* 99–106

Britton, D.K. (1977) 'Some explorations in the analysis of long-term changes in the structure of agriculture', *Journal of Agricultural Economics, 28,* 197–209

Brown, L.A. (1975) 'The market and infrastructure context of adoption: a spatial perspective on the diffusion of innovation', *Economic Geography, 51,* 185–216

Brown, L.A. (1981) *Innovation Diffusion: a New Perspective,* Methuen, London

Brown, L.A. and Lentnek, B. (1973) 'Innovation diffusion in a developing economy: a mesoscale view', *Economic Development and Cultural Change, 21,* 274–92

Brown, M.A. (1980) 'Attitudes and social categories: complementary explanations of innovation adoption', *Environment and Planning A, 12,* 175–86

Brown, M.A., Maxon, G.E. and Brown, L.A. (1977) 'Diffusion-agency strategy and innovation diffusion: a case study of the Eastern Ohio Resource Development Centre', *Regional Science Perspectives, 7,* 1–26

Bryant, C.R. (1974) 'An approach to the problem of urbanisation and structural change in agriculture: a case study from the Paris region, 1955–68', *Geografiska Annaler, 56B,* 1–27

Bryant, C.R. (1981) 'Agriculture in an urbanising environment: a case study from the Paris region, 1968–76', *Canadian Geographer, 25,* 27–45

Buchanan, K.B. (1948) 'Modern farming in the vale of Evesham', *Economic Geography, 24,* 235–50

Bunting, T. and Guelke, L. (1979) 'Behavioural and perception geography: a critical appraisal', *Annals of the Association of American Geographers, 69,* 448–63

Butler, J.B. (1960) *Profit and purpose in farming: a study of farms and smallholdings in part of the North Riding,* University of Leeds, Department of Economics, p. 68

Buttel, F. and Larson, O.W. (1979) 'Farm-size, structure and energy intensity: an ecological analysis of US agriculture', *Rural Sociology, 44,* 471–88

Carlyle, W. (1983) 'Farm lay-outs in Manitoba', *Canadian Geographer, 27,* 17–34

Chrisholm, M. (1979) *Rural Settlement and Land Use,* Hutchinson, London

Clark, G. (1979) 'Farm amalgamations in Scotland', *Scottish Geographical Magazine, 95,* 93–107

Clout, H.D. (1968) 'Planned and unplanned changes in French farm structures', *Geography, 53,* 311–15

Clout, H.D. (1972) *Geography of Post-war France: a Social and Economic Approach,* Pergamon Press, Oxford

Clout, H.D. (1975) 'Structural changes in French farming: the case of the Puys-de-Dome', *Tijdschrift voor Economische en Sociale Geografie, 66,* 234–45

Coleman, A. (1978) 'Agricultural land losses: the evidence from maps' in A.W. Rogers (ed.) *Urban growth, Farmland Losses and Planning,* Institute of British Geographers, Wye College

Coppock, J.T. (1964) 'Post-war studies in the geography of British agriculture', *Geographical Review, 54,* 409–26

Coppock, J.T. (1968) 'The geography of agriculture', *Journal of Agricultural Economics, 19,* 153–75

Cromley, R. (1982) 'The von Thünen model and evironmental uncertainty', *Annals of the Association of American Geographers, 73,* 404–10

Dalton, R.T. (1971) 'Peas for freezing: a recent development in Lincolnshire agriculture', *East Midlands Geographer, 5,* 133–41

Day, R. and Tinney, D. (1969) 'A dynamic von Thünen model', *Geographical Analysis, 1,* 137–51

De Garis De Lisle, D. (1982) 'Effects of distance on cropping patterns internal to the farm', *Annals of the Association of American Geographers, 72,* 88–98

Dennett, M.D., Elston, J. and Diego, Q.R. (1980) 'Weather and yields of tobacco, sugar-beet and wheat in Europe', *Agricultural Meteorology, 21,* 249–63

Edwards, A.M. and Wibberley, G.P. (1971) *An Agricultural Land Budget for Britain, 1965–2000,* Wye College

Edwards, C.J. (1978) 'The effects of changing farm size upon levels of farm fragmenta-tion: a Somerset case study', *Journal of Agricultural Economics, 29,* 143–54

Ewald, U. (1976) 'The von Thünen principle and agricultural zonation in colonial Mex-ico', *Journal of Historical Geography, 3,* 123–34

Floyd, B. (1976) *Problems in the modernisation of small-scale agriculture in underdeveloped tropical countries: a case study from the Caribbean,* XXII International Geographical Conference, USSR

Garst, R.D. (1974) 'Innovation diffusion among the Gusii of Kenya', *Economic Geography, 50,* 300–12

Gasson, R.M. (1973) 'Goals and values of farmers', *Journal of Agricultural Economics, 24,* 521–42

Gillooly, J.F. (1978) 'On the association of soil types and maize yields', *South African Journal of Science, 74,* 138–9

Gillooly, J.F. and Dyer, T. (1979) 'On spatial and temporal variations of maize yields over South Africa', *South African Geographical Journal, 61,* 111–18

Golledge, R.G. (1960) 'Sydney's metropolitan fringe: a study in rural-urban relations', *Australian Geographer, 7,* 243–55

Gould, P.R. (1963) 'Man against his environment: a game-theoretic framework', *Annals of the Association of American Geographers, 53,* 291–7

Granger, O. (1980a) 'Climatic variations and the Californian raisin industry', *Geographical Review, 70,* 300–13

Granger, O. (1980b) 'The impact of climatic variation on the yield of selected crops in three California counties', *Agricultural Meteorology, 22,* 367–86

Gregor, H.F. (1963) 'Urbanisation of south Californian agriculture' *Tijdschrift voor*

Economische en Sociale Geografie, *54*, 273-8

Gregor, H.F. (1979) 'The large farm as a stereotype: a look at the Pacific Southwest', *Economic Geography*, *55*, 71-87

Gregor, H.F. (1982) 'Large-scale farming as a cultural dilemma in US rural development — the role of capital', *Geoforum*, *13*, 1-10

Griffin, E. (1973) 'Testing von Thünen's theory in Uruguay', *Geographical Review*, *63*, 500-16

Grigg, D.B. (1982a) *The Dynamics of Agricultural Change*, Hutchinson, London

Grigg, D.B. (1982b) 'Agricultural geography', *Progress in Human Geography*, *6*, 242-6

Grigg, D.B. (1983) 'Agricultural geography', *Progress in Human Geography*, *7*, 255-60

Grigg, D.B. (1984) *An Introduction to Agricultural Geography*, Hutchinson, London

Hagerstrand, T. (1967) *Innovation Diffusion as a Spatial Process*, University of Chicago Press, Chicago

Haines, M.R. (1982) *An Introduction to Farming Systems*, Longmans, London

Hart, P.W.E. (1978) 'Geographical aspects of contract farming, with special reference to the supply of crops to processing plants', *Tijdschrift voor Economische en Sociale Geografie*, *69*, 205-15.

Hart, P.W.E. (1980) 'Problems and potentialities of the behavioural approach to agricultural location', *Geografiska Annaler*, *62B*, 99-108

Harvey, D.W. (1963) 'Locational change in the Kentish hop industry and the analysis of land-use patterns', *Transactions of the Institute of British Geographers*, *33*, 123-44

Harvey, D.W. (1966) 'Theoretical concepts and the analysis of agricultural land use patterns in geography', *Annals of the Association of American Geographers*, *36*, 362-74

Havens, A.E. and Flinn, W.L. (1975) 'Green revolution technology and community development: the limits of action programmes', *Economic Development and Cultural Change*, *23*, 469-81

Hewlett, R. (1967) 'Status, achievements and problems of agricultural co-operatives in Europe' in T.K. Warley (ed.), *Agricultural Producers and Their Markets*, Blackwell, Oxford

Hill, R. and Smith, D.L. (1977) 'Farm fragmentation on western Eyre Peninsula, South Australia', *Australian Geographical Studies*, *15*, 158-73

Horvath, R.J. (1969) 'Von Thünen's isolated state and the area around Addis Ababa, Ethiopia', *Annals of the Association of American Geographers*, *59*, 308-23

Ilbery, B.W. (1977) 'Point score analysis: a methodological framework for analysing the decision-making process in agriculture', *Tijdschrift voor Economische en Sociale Geografie*, *68*, 66-71

Ilbery, B.W. (1978) 'Agricultural decision-making: a behavioural perspective', *Progress in Human Geography*, *2*, 448-66

Ilbery, B.W. (1979) 'Decision-making in agriculture: a case study of north-east Oxfordshire', *Regional Studies*, *13*, 199-210

Ilbery, B.W. (1983) 'Goals and values of hop farmers', *Transactions of the Institute of British Geographers*, *8*, 329-41

Ilbery, B.W. (1984a) 'Farm fragmentation in the Vale of Evesham', *Area*, *16*, 159-65

Ilbery, B.W. (1984b) 'Agricultural specialisation and farmer decision behaviour in the West Midlands', *Tijdschrift voor Economische en Sociale Geografie*, *75*, 329-34

Ilbery, B.W. (1985) *Agricultural Geography: a Social and Economic Analysis*, Oxford University Press, Oxford

Ingersent, K.A. (1979) 'The variability of British potato yields: a statistical analysis', *Oxford Agrarian Studies*, *8*, 33-52

Johansen, H.E. (1971) 'Diffusion of strip cropping in southwestern Wisconsin', *Annals of the Association of American Geographers*, *61*, 671-83

Jones, G.E. (1967) 'The adoption and diffusion of agricultural practices', *World Agricultural Economics and Rural Sociology Abstracts*, *9*, 1-34

Jones, R.C. (1976) 'Testing macro-Thünen models by linear programming', *Professional*

Geographer, 28, 353-61

Kampp, A, (1979) 'Recent amalgamation of agricultural holdings', *Geografisk Tidsskrift*, 78, 57-60

Kellerman, A. (1977) 'The pertinence of the macro-Thünian analysis', *Economic Geography*, 53, 255-64

King, R.L. (1977) *Land Reform: a World Survey*, Bell, London

King, R.L. and Burton, S. (1982) 'Land fragmentation: notes on a fundamental rural spatial problem', *Progress in Human Geography*, 6, 476-94

King, R.L. and Burton, S. (1983) 'Structural change in agriculture: the geography of land consolidation', *Progress in Human Geography*, 7, 471-501

Klages, K.H.W. (1942) *Ecological Crop Geography*, Macmillan, New York

Kreuger, R.R. (1978) 'Urbanisation of the Niagara fruit belt', *Canadian Geographer*, 22, 179-93

Layton, R.L. (1981) 'Attitudes of hobby and commercial farmers in the rural-urban fringe of London, Ontario', *Cambria*, 8, 33-44

Le Vay, C. (1983) 'Agricultural co-operative theory: a review', *Journal of Agricultural Economics*, 34, 1-44

Linstrom, H.R. (1978) *Farmer to consumer marketing*, USDA, Economic Research Service, Washington, DC

Mattingley, P.F. (1972) 'Intensity of agricultural land-use near cities: a case study', *Professional Geographer*, 24, 7-10

Merrill, R. (ed.) (1976) *Radical Agriculture*, Harper and Row, London

Michaels, P. (1982) 'Atmospheric pressure patterns, climatic change and winter wheat yields in North America', *Geoforum*, 13, 263-73

Misra, R.P. (1969) 'Monte Carlo simulation of spatial diffusion: rationale and application to the Indian condition' in R.P. Misra (ed.), *Regional Planning*, University of Mysore Press, Mysore

Moran, W. (1979) 'Spatial patterns of agriculture on the urban periphery: the Auckland case', *Tijdschrift voor Economische en Sociale Geografie*, 70, 164-76

Morgan, W.B. and Munton, R.J.C. (1971) *Agricultural Geography*, Methuen, London

Muller, P. (1973) 'Trend surfaces of American agriculture: a macro-Thünen analysis', *Economic Geography*, 49, 228-42

Naylor, E.L. (1982) 'Retirement policy in French agriculture', *Journal of Agricultural Economics*, 33, 25-36

Noble, G. (1980) 'Farm management problems of the energy crisis', *ADAS Quarterly Review*, 36, 1-13

Norton, W. (1979) 'Relevance of von Thünen theory to historical and evolutionary analysis of agricultural land use', *Journal of Agricultural Economics*, 30, 39-47

Parry, M.L. (1975) 'Secular climatic change and marginal agriculture', *Transactions of the Institute of British Geographers*, 64, 1-14

Parry, M.L. (1976) 'Mapping of abandoned farmland in upland Britain', *Geographical Journal*, 142, 101-10

Peet, J.R. (1969) 'The spatial expression of commercial agriculture in the nineteenth century: a von Thünen interpretation', *Economic Geography*, 45, 283-301

Platt, R. (1977) 'The loss of farmland: evolution of public response', *Geographical Review*, 67, 93-101

Ricardo, D. (1817) *Principles of Political Economy and Taxation*, Dent, London

Richardson, B. (1974) 'Distance regularities in Guyanese rice cultivation', *Journal of Developing Areas*, 8, 235-55

Rogers, E.M. and Shoemaker, F.F. (1971) *Communication of Innovations: a Cross-cultural Approach*, Free Press, New York

Simmons, I.G. (1979) *Biogeography, Natural and Cultural*, Edward Arnold, London

Simmons, I.G. (1980) 'Ecological-functional approaches to agriculture in geographical contexts', *Geography*, 65, 305-16

Sinclair, R.J. (1967) 'Von Thünen and urban sprawl', *Annals of the Association of American*

Geographers, 57, 72–87

Sit, V.F.S. (1979) 'Agriculture in urban shadow: a review of the post-war experience of Hong Kong', *Pacific Viewpoint, 20,* 199–209

Smith, E.G. (1975) 'Fragmented farms in the USA', *Annals of the Association of American Geographers, 65,* 58–70

Smith, E.G. (1980) 'America's richest farms and ranches', *Annals of the Association of American Geographers, 70,* 528–41

Symons, L.J. (1978) *Agricultural Geography,* Bell, London

Tarrant, J.R. (1974) *Agricultural Geography,* David and Charles, Newton Abbot

Tavener, L.E. (1952) 'Changes in the agricultural geography of Dorset, 1929–49', *Transactions of the Institute of British Geographers, 18,* 93–106

Todd, D. (1979) 'Regional and structural factors in farm-size variations: a Manitoba elucidation', *Environment and Planning A, 11,* 237–58

Townsend, J.G. (1977) 'Perceived worlds of the colonists of tropical rainforest, Columbia', *Transactions of the Institute of British Geographers, 2,* 430–58

Van Valkenburg, S. and Held, C.C. (1952) *Europe,* Wiley, London

Varjo, U. (1979) 'Productivity and fluctuating limits of crop cultivation in Finland', *Geographica Polonica, 40,* 225–33

Vermeer, D.E. (1981) 'Collision of climate, cattle and culture in Mauritania during the 1970s', *Geographical Review, 71,* 281–97

von Thünen, J.H. (1826) *Der isolierte Staat in Beziebung auf Ladwirtschaft und Nationalökonomie,* Rostock

Wallace, I. (1985) 'Towards a geography of agribusiness', (unpublished manuscript)

White, R.L. and Watts, H.D. (1977) 'The spatial evolution of an industry: the example of broiler production', *Transactions of the Institute of British Geographers, 2,* 175–91

Whittlesey, D. (1936) 'Major agricultural regions of the earth', *Annals of the Association of American Geographers, 26,* 199–240

Winsberg, M. (1980) 'Concentration and specialisation in US agriculture, 1939–78, *Economic Geography, 56,* 183–9

Wolpert, J. (1964) 'The decision-making process in a spatial context', *Annals of the Association of American Geographers, 54,* 537–58

Wood, L.J. (1981) 'Energy and agriculture: some geographical implications' *Tijdschrift voor Economische en Sociale Geografie, 72,* 224–34

Yapa, L.S. and Mayfield, R.C. (1978), 'Non-adoption of innovations: evidence from discriminant analysis', *Economic Geography, 54,* 145–56

2 CLASSIFICATION OF AGRICULTURAL SYSTEMS

J.W. Aitchison

Introduction

Agricultural geographers have a long-standing and deep-seated interest in the identification, classification and regionalisation of farming systems (Bonnamour, 1973; Gregor, 1970; Grigg, 1969; McCarty, 1954; Reeds, 1964; Whittlesey, 1936). Indeed it could be argued that this taxonomic perspective constitutes the most abiding of all the research traditions within the discipline. Over the past two decades, despite provocative criticisms from several quarters (e.g. Chisholm, 1964; Hart, 1975; Morgan and Munton, 1971; Spencer and Stewart, 1973), it has continued to maintain a firm, if less pervasive, intellectual hold. Given the innate complexity of agricultural systems, their variety and protean nature, it is perhaps not too surprising that so many geographers should still feel moved to structure typologies and to explore associated patterns of regional variation. Of course, on a broader philosophical and methodological front, agricultural geographers will always have cause to engage in classificatory exercises, even though some exercises might not be central to the goals of particular research investigations. This is because classification constitutes a key element in all scientific inquiry. Whatever the problem and whatever the methodological stance, it is difficult to envisage a situation in which it is not necessary at some stage or other to put order on the objects, processes or events that are being analysed. Without some form of classification, however rudimentary, there can be no description or explanation of either patterns or processes; likewise in an applied planning context there can be no prescription. This said, however, it has to be recognised that to classify is all too easy; to classify well requires a sensitive and practical understanding not only of the systems under investigation, but also of what is involved in the classificatory process itself — the alternative strategies that might be adopted and the often subjective decisions that have to be taken. Classification is a 'boundedly objective decision-making process' (Aitchison, 1975, p. 17), and will always be so.

Figure 2.1 seeks to define in simplified diagrammatic form the main phases and conceptual issues within the classificatory process. Here, four

Figure 2.1: Stages in the Process of Classification

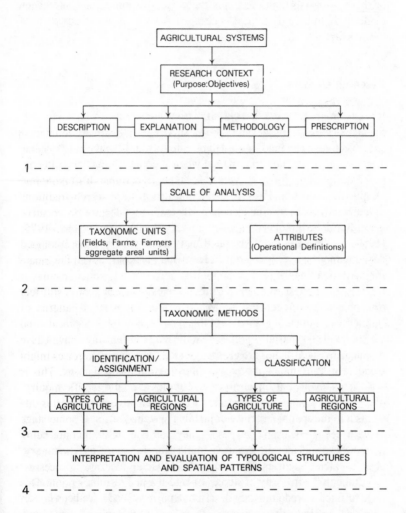

separate but integrated stages are recognised. The first refers to the *research context* within which the proposed classification is set; the second identifies a *design stage* in which the taxonomic units are identified, together with the attributes to be employed in the classification; the third recognises an *analytic stage* in which classificatory measures and methods are selected; finally, there is an *interpretation and evaluation stage* in which the resultant typology is assessed in the light of the goals of the inquiry. This review outlines the essential nature of each of these four stages and assesses the varied responses that have been made to them by agricultural geographers working in various parts of the world.

Research Context

It is customary to stress that classifications are structured to satisfy particular purposes (Grigg, 1969). For a successful classification the aims and objectives of an inquiry need to be carefully and explicitly articulated. They are of paramount importance when it comes to making decisions at the design, analytic and evaluative stages of a research programme. Despite this, Tarrant (1974, p.104) has quite rightly claimed that 'clarity of purpose has been conspicuously absent' from the works of many agricultural geographers engaged in classificatory inquiries. Harvey (1969, p. 326), has likewise stated that 'classifications have been produced without it ever being quite clear what purposes they are designed for'. All too often the resultant classifications (most commonly of farming regions) appear to be ends in themselves, satisfying mere *descriptive* objectives. According to Morgan and Munton (1971, p.104), 'immense classificatory labours' of this type have 'brought small results'. Chisholm (1964) has also argued that the maps of farming-type regions produced by such studies are generally of limited value and, like icebergs, frequently hide much more than they reveal. Whilst many of the arguments levelled at 'pattern-driven' studies are valid, it has to be emphasised that taxonomic exercises of this type can yield interesting insights into the spatial structure of farming systems, and at the same time serve a deeper heuristic purpose. They can trigger the generation of hypotheses, and thereby create a foundation for more profound 'process oriented' inquiries (Anderson, 1975; Grigg, 1969).

Although the majority of studies of land-use and farming-type regions have fulfilled a predominantly descriptive purpose, and have been largely empirical and inductive in nature, others have been explicitly tied to particular models of explanation. The array of studies structured to test the applicability of von Thünen's partial equilibrium model, for instance, have inevitably demanded categorisations of land-use and types of farming, with a view to testing the significance of the distance factor — via its effect on the cost of transporting both inputs and outputs — in shaping spatial variations in patterns of agricultural production (Blaikie, 1971; Chisholm, 1979). In similar deductive vein, Boserup has formulated a typology of systems of supply for vegetable food in order to demonstrate the relationship between population densities in subsistence farming areas and the general process of agricultural change (Boserup, 1965, 1981). Based on cropping frequencies her typology identifies six categories of vegetable supply systems (food gathering, forest-fallow, bush-fallow, short-fallow, annual cropping and multicropping). Less theoretically formulated, numerous studies have derived classifications in order to highlight associations between types of farming and other factors of production. The influence of various facets of the physical environment on land use and enterprise structures has attracted interest over the years (e.g. Aitchison, 1968, 1972; Cruickshank and Armstrong, 1971; Dennett, Elston and Speed, 1981; Grigg, 1970, 1984; Hidore, 1965; Munton, 1972; Taylor, 1952; Varjo, 1977), whilst more recently attention has been directed towards the development of typologies based on the behavioural characteristics of individual farmers and growers (e.g. Aitchison and Aubrey 1982; Bryant, 1981; Fletcher, 1983; Ilbery 1983a,b).

Although classificatory studies in agricultural geography have been overwhelmingly concerned with the description and explanation of regional variations in types of production, an increasingly important, and often associated, objective has been to explore more general methodological issues of relevance at the design and analytic stages of the taxonomic process. Particularly noteworthy in this regard have been efforts to demonstrate alternative statistical approaches to the derivation of agricultural typologies and regionalisations. It is with developments of this type that the major part of this discussion is to be concerned.

Unlike agricultural economists, agricultural geographers have not tended to structure typologies of farms and farming regions with applied or policy purposes in mind. This is not to imply that the work that has been carried out is all simply of academic interest. World-wide, much of it is used or cited by agencies and organisations specifically concerned with the management and development of the agricultural industry. Numerous papers presented at conferences of the International Geographical Union: Commission on Agricultural Typology — and the interest shown in them by the Food and Agriculture Organization — clearly indicate that this is so. In general, however, it could be argued that agricultural geographers have been excessively concerned with the regionalisation of types of land-use or types of farming, and that the compass of taxonomic inquiry might be rewardingly broadened with a view to satisfying more practical and prescriptive ends. There has been some progress on this front, but there is clearly considerable scope for typological analyses that focus on problems associated with particular farming systems or communities (e.g. typologies of marginal or deprived farms; behavioural and socio-political typologies of farmers relating to such matters as levels of access to land, information, technology and credit). To suggest a stronger reorientation towards more policy- or management-relevant typologies does not mean that more academic (geographical) objectives should be eschewed (e.g. the integrated development of spatial typologies and models of explanation), or that the heavy traditional emphasis on regionalisation is fruitless. The point being made is that the classificatory and typological tradition within agricultural geography could be enriched and rejuvenated by adoption of a more applied stance (Bowman, 1932; Gregor, 1970, pp. 13–16). A not unrelated matter, and one which has been stressed by others (Chisholm, 1964; Spencer and Stewart, 1973), is that agricultural geographers have also tended, in focusing upon the spatial dimension, to ignore the dynamic and evolving character of agricultural systems, and the socio-economic and ecological forces that are responsible for the changes that have been or are taking place within a particular regional setting.

Finally, it should perhaps be emphasised that the four 'purposes' recognized in Figure 2.1 are not necessarily discrete or separate entities. In practice most classificatory studies seek to satisfy several objectives. Papers by agricultural geographers commonly constitute a blend of description, explanation and methodology — sometimes with a dash of applied comment.

Design Stage

The process of classification is a search for pattern and order within a specified set of data. Data sets can be viewed as matrices of attribute scores for collections of taxonomic units (i.e. the objects to be classified). Clearly, the number of objects and attributes that make up such matrices, as well as their substantive nature, should reflect the objectives of a particular inquiry. The size of the matrix is important for it determines the 'scale' of the analysis — its level of resolution. As in all geographical inquiry, the scale factor not only affects the form of resultant patterns, it also influences interpretations of these patterns (Harvey, 1968). As Harvey (1969, p. 384) has noted 'inferences as to process derived from pattern analysis are not independent of the scale of analysis'. Whilst the logic of the situation demands that the scale issue be resolved by reference to the purposes of the inquiry, in practice agricultural geographers often find themselves constrained by the nature of available data. Indeed, such a strong influence has it had on the direction, quality and utility of classificatory research that it is possible to talk of the 'tyranny' of available data. In many instances it seeems as if the data dictate the purpose, rather than the reverse.

Traditionally, the majority of agricultural geographers have classified in order to regionalise. Since regionalisation is essentially 'areal' classification, it is not surprising that the taxonomic units used have dominantly been segments of agricultural space. These segments vary in size and shape, and some are more artificial (agriculturally speaking) than others. Thus, studies of land-use patterns generally take 'fields' as their unit of taxonomic reference. Investigations of this type (national land-use surveys apart) have been obliged to confine themselves to relatively small areas. With the increasing sophistication and availability of remote sensing systems, however, it is evident much wider vistas will be opened up for those concerned with land-use mapping and with land-use change at this 'field' level of resolution (Tarrant, 1974).

Within the body of classificatory literature the most interesting and meaningful studies are those that have taken the 'farm' as their unit of taxonomic reference. This is because farms — using the term in its broadest possible sense — are bounded systems, the character and functioning of which derive from decisions taken by individuals or groups of individuals. In short, the farm is the natural organisational domain of the agricultural landscape. To a large extent it is the variable resource endowment of farms, set against the needs and aspirations of those managing them, that ensures diversity of structure and practice within a given region; it is precisely this diversity that agricultural geographers seek to describe and understand. Unfortunately, in many parts of the world, data for farms (agricultural holdings) are seldom readily available. The would-be taxonomist is therefore generally obliged to garner information at farm level through questionnaire-based field surveys (Birch, 1954). Resource limitations inevitably constrain the scope of such surveys, and it is partly for this reason that typologies of farms are less frequently encountered in the published literature than might otherwise have been expected. This does not mean that farm studies have failed to attract attention in the broader academic sphere. The number of commendable, but none the less forgotten, farm studies carried out by undergraduates and post-graduates throughout the world for dissertations must be enormous.

Despite all the problems that arise in interpreting aggregated statistics (Coppock, 1960; Harvey, 1969; Weaver, 1956) agricultural geographers have continued to rely heavily upon information which may have been collected at farm level (e.g. by government agencies), but which for convenience (or to ensure confidentiality of individual farm returns) is collated and published for higher-order areal units. These units are generally administrative structures (e.g. parishes, communes, counties, provinces or states), the boundaries of which are often quite meaningless — at least as far as the agricultural geography of the reference region is concerned. Furthermore, since these units are seldom of standard size or shape, descriptive statistics derived from aggregate data (e.g. ratios, densities, means, measures of dispersion) need to be treated with considerable circumspection. It cannot be assumed that they constitute a uniformly representative set of data, on the basis of which reliable comparisons can be made or inferences drawn. To overcome this problem it has been suggested that, where necessary, attempts should be made to create more standardised areal units, either by a process of further aggregation (e.g. combining sets of parishes with a view to creating a collection of more uniform divisions — Aitchison, 1979; Coppock, 1960) or by collating data within regular grids such as squares or hexagons (van Hecke, 1983). It is not necessary here to elaborate upon the various statistical inadequacies of aggregated data or on the inferential difficulties that apply. One issue worthy of mention, however, is the distinction between 'singular' and 'collective' forms of statistical aggregation. Where the aggregate information maintains a reference to some lower-order spatial unit (e.g. numbers of farms of particular types), the area concerned should be conceived of as a 'collective' entity. The same area becomes a 'singular' entity, however, when the information refers to the total extent or magnitude of a particular attribute (e.g. acreages of particular crops, numbers of particular types of livestock). With more and more agricultural censuses providing both types of data, and with both being amenable to classificatory analysis, this difference is of some import for it poses 'conceptual and inferential problems' (Harvey, 1969, p. 352) and demands close logical scrutiny, especially when patterns of spatial covariance are under investigation (Aitchison, 1980; Talman, 1979).

A further issue that can complicate endeavours to structure agricultural typologies or to carry out regionalisations, and one which applies most particularly to farm-based studies, is that of deriving appropriate and representative samples of taxonomic units. Once again the manner in which samples are drawn (e.g. purposive sampling, spatial sampling sytems, stratified random sampling) and their size should be directly related to the purpose of the investigation (Belshaw and Jackson, 1966; Berry, 1962; Birch, 1954; Blaut, 1959; Board, 1970; Haggett, 1963;

Wood, 1955). It is to be expected, for instance, that sampling strategies selected for more general typological studies would differ from those where the main interest is in regionalisation. In the former case stratification on the basis of particular internal attributes may be appropriate (e.g. farm size, land quality), whilst in the latter the main concern may be to ensure a spatially representative sample.

Of key significance within the design stage of the classificatory process is the selection and operational definition of the attributes to be used in differentiating and categorising the systems under investigation. The debates that such matters have generated over the years bear testimony to their importance. Bridging, as they do, the interface between physical and human domains, it is to be expected that farming systems can be characterised by reference to a manifold series of attributes. The would-be taxonomist must once more exercise judgement in selecting those attributes which satisfy the aims of a particular inquiry. Since the delimitation of farming-type regions has long constituted a major focus of research interest it is perhaps not surprising that the bulk of studies should have chosen to structure classifications using statistics relating to levels of crop and livestock production. Whilst many early taxonomies were concerned to describe patterns of spatial variation in the production of single agricultural commodities or enterprises (Gregor, 1970), a more dominant tendency has been to derive multivariate classifications in which crop and livestock systems are treated in an integrated manner. To achieve this integration various solutions have been effected. The most popular approach has been to convert straight area figures for crops and headage figures for livestock (the form in which most census information is normally presented) to scales of measurement that are either more revealing or that facilitate attribute comparisons. Within the literature are to be encountered numerous examples of the use of different types of conversion coefficients. Studies of pastoral farming systems have weighted livestock of different types according to their standard feed requirements, whilst efforts to place both crops and livestock on common measurement scales have led to the use of coefficients based on standard labour requirements (e.g. standard man-days), grain or wheat equivalents, and such monetary measures as standard gross margins and gross outputs. Although conversion solutions of this type are useful in assessing the relative significance of particular enterprises and when using taxonomic methods that operate on closed systems of percentages, the conversion coefficients themselves are open to various criticisms (Jenkins, 1982; Morgan and Munton, 1971). However, since few data collection agencies publish actual farm data in this form — actual as opposed to standardised — conversion coefficients will continue to serve an important purpose. It also has to be accepted that these generalised coefficients inevitably become cruder as the area of regional reference increases and, furthermore, that they are not stable over time. This clearly bedevils efforts at comparative analysis. Even within a small area such as the British Isles, for instance, agronomists have established different sets of livestock equivalent units to suit different regional environments. The coarseness of the animal and grain equivalent units adopted by the IGU Commission on Agricultural Typology and the regionally differentiated gross margin figures adopted by the European Community in establishing a typology of holdings afford graphic illustrations of the problems involved and the compromises that have to be made. It also has to be appreciated that the process of converting crop and livestock statistics to common scales does not necessarily mean that the taxonomist is in a position to classify farming systems according to their functional structures. As Chisholm (1964) has noted, many studies fail to isolate functional linkages between the various facets of production at the farm level (e.g. the extent to which cropping systems are discrete enterprises in their own right or constitute sources of fodder for on-farm livestock). Data on such linkages are notoriously difficult to come by. Agricultural censuses commonly collate data for items of production without endeavouring to indicate their roles in the total enterprise system.

Although crop and livestock characteristics figure strongly in the majority of studies, agricultural geographers have in fact developed classifications based on a whole variety of attributes. The physical and biological endowments of farms and farming areas have long been used in structuring classifications, but whether or not such factors of production are appropriately included in studies of types of farming regions has been queried by several writers. Birch (1954), Buchanan (1959), Chisholm (1964), Grigg (1969), and Whittlesey (1936), for instance, have all criticised the classic works of Baker (1926–32) and certain of his contemporaries for the way in which they delimited the boundaries of agricultural regions (e.g. the 'Cotton' and 'Corn' belts) by searching for edaphic, climatic or physiographic breaks of slope to which changes in types of farming practice could be allied. Whilst the quality of the regionalisation strategies adopted and the soundness of the inferences drawn in a number of these early studies are clearly questionable, this does not mean that attributes pertaining to the physical resource endowment of farming systems should be avoided in building typologies. Munton (1972), in a detailed study of system relationships, ordinated a sample of 218 farms in England by reference to 80 attributes. Of these, 31 were environmental in nature (e.g. the texture, depth, acidity and workability of soils). Similarly, in investigations into the structure of peasant agriculture in Barbados, Henshall (1966) has included such attributes as soil type, slope, depth of soil, stoniness and levels of erosion. With a view to testing hypotheses proposed by Taylor (1952), Aitchison (1968) derived two separate classifications of holdings within the arable farming region of south-west Lancashire — the one based on the areal dominance of particular soil types and the other on cropping systems. The objective here was to assess the statistical significance of the relationship between these two classificatory structures.

Besides environomental variables, agricultural geographers have commonly included a range of structural and input attributes in the classification of farms and farming areas. Too diverse to list, the characteristics and measures that have been used embrace such items as size of holding in terms of land, capital and labour, field patterns, crop rotation systems, farm building types, tenurial conditions, levels of mechanisation, distance to markets, degrees of commercialisation, etc. Partly because of problems of data acquisition and partly because of the conceptual difficulties involved, geographers have been rather slow in developing typologies based on the personal behavioural attributes of farmers and growers. Given the importance of such matters when it comes to explaining patterns of spatial variation in, say, types of farming and their relevance in many practical contexts (e.g. in seeking to implement agricultural development programmes), this is clearly an area of study worthy of further consideration. Ilbery (1983a), for instance, has suggested that the work of Gasson (1973) on the goals and values of farmers is of particular interest in this regard. Conceptualising the problem at a higher 'structuralist' level Spencer and Stewart (1973) and Andrianov and Cheboksarov (1975) have argued that in developing typologies more attention should be paid to cultural and historico-ethnographic attributes.

Analytic Stage

Having structured an appropriate data set the would-be taxonomist is now faced with the problem of how to analyse it. In practice of course this problem would normally have been resolved even before the information-gathering stage. This is because the level at which attributes are measured (e.g. categorical, ordinal or numeric) is of importance when it comes to selecting a particular analytic strategy. Classificatory studies in agricultural geography have in fact used almost every conceivable form of measurement and statistical expression, from simple binary records to orthonormalised scores derived from various types of factor analysis. This variety of measurement forms complements an equally diverse collection of taxonomic methods.

In Figure 2.1 it is proposed that the taxonomic methods adopted by agricultural geographers are of two basic types. The first includes what statisticians refer to as methods of '*identification*' or '*assignment*'; the second are generally referred to as methods of '*classification*'.

Methods of Identification

The distinctive feature of these strategies is that the taxonomist sets up an *a priori* group structure and then proceeds to assign individual taxonomic units to those groups with which they are most closely identified. The 'least squares' procedure devised by Weaver (1954a, b), and subsequently modified slightly by Thomas (1963), is a method of assignment that has proved to be particularly popular. In the history of typological research in agricultural geography it constitutes a major methodological milestone. Previous efforts to distinguish types and regions of agriculture were often highly subjective in approach. Here, for the first time was a controlled, objective procedure eminently suited to the analysis of percentage data. The method is undoubtedly limited in its applicability (based as it is on closed number sets) and can lead to messy solutions (in the sense that it often generates large numbers of typal combinations and permutations), but be that as it may, it has been, and continues to be, successfully employed in the identification of types of farms and types-of-farming areas (Aitchison, 1980; Coppock, 1964a; Gillmor, 1977; Scott, 1957). The *a priori* groups recognised within the least-squares method are theoretical ideals which indicate degrees and types of specialisation or diversification. The taxonomist simply calculates deviations between an observed profile of proportions (e.g. percentages of a total farm area under different categories of land-use) and an ordered suite of ideal profiles (e.g. from a profile where only one type of land-use is recorded, through a graded series, to one in which all types of land-use are present in equal proportions). The minimum deviation identifies the ideal profile to which the taxonomic unit is to be assigned. Stylistically similar are assignment strategies which use reference profiles that are empirical rather than theoretical. Perpillou (1952, 1970, 1977), for instance, has categorised types of land-use in France at the commune level by relating observed proportions to '*un terroir de référence*' (i.e. the profile of land-use proportions for France as a whole). Guermond and Massias (1973) have compared the relative merits of this approach and that proposed by Weaver.

A less rigorous identification procedure involves the use of ternary diagrams. Although only allowing reference to three attributes, these graphic devices have been widely used in assigning farms and farming areas to typal groups. Ternary diagrams are normally divided into segments by lines which mark percentage threshold values for particular attributes. The boundaries of these segments constitute class limits, and each individual taxonomic unit is identified according to the segment into which it falls. The problem here is how to select appropriate class boundaries. These may be theoretically determined (Aitchison, 1981; Ingram, 1984), arbitrarily selected, or derived empirically (e.g. by searching for 'natural' breaks within the data set). In a study of farming patterns in East Anglia, Jackson, Barnard and Sturrock (1963) used ternary diagrams to identify different types of cropping and livestock farms. The three enterprise categories employed in identifying types of cropping farms were grain, roots and horticulture. For livestock farms the three categories were dairying, cattle rearing and fattening, and sheep. The relative significance of enterprises was determined using standard gross outputs. Since plots of farms within the ternary diagrams revealed no 'natural' groupings, typal classes were identified using arbitrary 20 per cent thresholds. Varjo (1977, 1984), in an analysis of Finnish farming systems, has also used ternary diagrams to identify seven basic enterprise combinations. In this case the three enterprises selected for study — arable farming, cattle rearing and forestry — are rated on the basis of gross margin contributions. To sectionalise the ternary diagram (Figure 2.2) Varjo takes the proportion of gross margins associated with each enterprise at national level. These are 45 per cent, 30 per cent and 25 per cent for forestry, cattle rearing and arable respectively. These thresholds yield six basic groups. A seventh group is inserted with a view to distinguishing those areas which are similar to the overall national set of proportions. Changes in enterprise structures are also considered by charting shifts in ternary locations for the years 1969 and 1975.

Agricultural data-collecting agencies in many countries frequently categorise farms by allocating them to pre-defined classes. Taxonomic frameworks may differ but the process involved is generally one of assignment. An example of such typologies is that established in 1978 by the European Economic Community (Commision of the European Communities, 1978, 1984). Using sets of regionally differentiated gross margin statistics for over 60 types of crops and livestock, farms are allocated to groups according to the proportionate dominance of a specified selection of enterprise categories. The typology itself is

Figure 2.2: Types of Farming in Finland

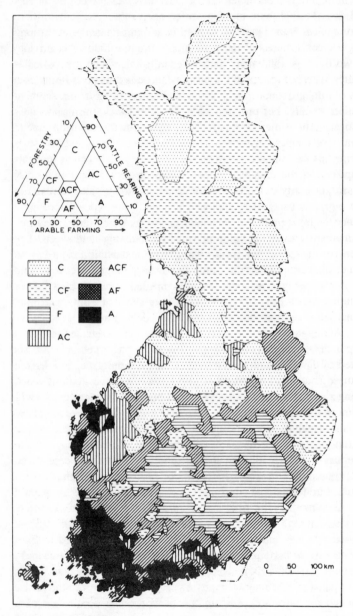

Source: Varjo, 1984.

hierarchically structured (Table 2.1) and recognises 17 'principal' types of farming and 58 'particular' types. Within the latter are to be identified 'specialist' systems (with specific enterprises accounting for more than two-thirds of total gross margins); 'partially dominant' systems (with a single specific cluster of enterprises accounting for between one-third and two-thirds of total gross margins) and 'bipolar' systems (with two specific clusters of enterprises accounting for between one-third and two-thirds of total gross margins). Whilst the main thresholds for type identification are 33.3 per cent and 66.6 per cent, in certain instances use is also made of 10 per cent and 25 per cent boundaries. The typology is clearly elementary in form, and criticisms have been levelled at the crudeness of the standard gross margin coefficients — even though they are updated to suit changing price and cost structure (Jenkins, 1982). This said, however, it has to be appreciated that the typology has to embrace a great variety of system types.

Since the latter part of the nineteenth century, numerous efforts have been made to structure agricultural typologies that are sufficiently comprehensive to capture variations in systems of production at global and continental scales. Given the diversity of the systems under investigation, many writers have chosen to formulate highly selective and qualitative typologies for purposes of generalised description and regionalisation (Gregor, 1970; Grigg, 1969, 1974; Spencer and Stewart, 1973). Although easy to criticise, such typologies are often founded upon a deep understanding of the systems concerned. The typological scheme adopted by Duckham and Masefield (1970), for instance, is a loosely structured, but for their purposes appropriate, classification of world farming systems based on four enterprise types (tree crops, tillage with or without livestock, alternating tillage with grass, bush or forest, and grassland), with further subdivisions according to intensity of production (an extensive-intensive continuum) and climate (tropical and temperate). In sharp conceptual contrast Andrianov and Cheboksarov (1972) recognise eight broadly-defined 'economic cultural' types of world agriculture on the basis of which they identify 34 'historical-ethnographic' areas (Andrianov, 1979). Analysing tropical farming systems Ruthenberg (1971) isolates nine *a priori* types between which, he freely admits, 'there are often no clear-cut divisions' (p. 282). The nine types include six cultivation systems (shifting cultivation, semi-permanent cultivation, regulated ley farming, permanent rain-fed cultivation, arable irrigation and perennial crops) and three grazing systems (total nomadism, semi-nomadism and ranching). Fully aware of the inadequacies of available data and sensitive to the true complexities of agriculture in the Third

Table 2.1: Typology of Agricultural Holdings: European Economic Community

Principal Types Code Heading		Particular Types Code Heading	
11	Cereals	111	Cereals, excluding rice
		112	Rice
		113	Cereals, including rice
12	Field crops, other	121	Roots
		122	Cereals and roots
		123	Field crops, various (*)
21	Horticulture	211	Market garden vegetables, open air
		212	Market garden vegetables, under glass
		213	Market garden vegetables, open air/under glass
		214	Flowers, open air
		215	Flowers, under glass
		216	Flowers, open air/under glass
		217	Horticulture, various (**)
31	Vineyards	311	Quality wine
		312	Table wine
		313	Table grapes
		314	Vineyards, mixed
32	Fruit/permanent crops, other	321	Fruit, excluding citrus
		322	Citrus
		323	Olives
		324	Permanent crops, various
41	Cattle, dairying	411	Specialized dairying
		412	Dairying, other

Field crops

Horticulture

Permanent crops

Grazing livestock			
	42	Cattle, rearing/fattening	421 Cattle, rearing/fattening, suckling
			422 Cattle, rearing/fattening, other
	43	Cattle, mixed	431 Dairying with cattle rearing/fattening
			432 Cattle rearing/fattening with dairying
			441 Sheep
	44	Grazing livestock, other	442 Cattle and sheep
			443 Grazing livestock, various
Pigs and poultry			
	51	Pigs	511 Pigs, rearing
			512 Pigs, fattening
			513 Pigs, mixed
	52	Pigs and poultry, other	521 Laying hens
			522 Table fowl
			523 Pigs and poultry, combined
			524 Pigs and poultry, various
	61	Horticulture and permanent crops	611 Horticulture and permanent crops
Mixed cropping			
			621 Field crops and horticulture
			622 Field crops and vineyards
			623 Field crops and fruit/permanent crops, other
	62	Mixed cropping, other	624 Partially dominant field crops
			625 Partially dominant horticulture or permanent crops
	71	Partially dominant grazing livestock	711 Partially dominant dairying
			712 Partially dominant grazing livestock other than dairying
Mixed livestock			
			721 Pigs and poultry and dairying
	72	Mixed livestock, other	722 Pigs and poultry and grazing livestock other than dairying
			723 Partially dominant pigs and poultry

Table 2.1 contd.

Principal Types		Particular Types	
Code	Heading	Code	Heading
81	Field crops and grazing livestock		
		811	Field crops with dairying
		812	Dairying with field crops
		813	Field crops with grazing livestock other than dairying
		814	Grazing livestock other than dairying with field crops
	Crops — livestock		
82	Crops — livestock, other		
		821	Field crops and pigs and poultry
		822	Crops — livestock, various

(*) Heading 123 'Field crops, various' is subdivided as follows if specifically required:
123 Field crops, various
    1231 Open field vegetables
    1232 Field crops, various, other

(**) Heading 217 'Horticulture, various' is subdivided as follows if specifically required:
    2171 Market garden vegetables, flowers, open air
    2172 Market garden vegetables, flowers, under glass
217 Horticulture, various
    2173 Mushrooms
    2174 Horticulture, mixed

World, Morgan (1977, pp. 170–220) likewise makes it clear that his typology (based on types of enterprises and market orientations) unavoidably ignores 'a great mass of "in-between" situations' (p. 174).

In 1964 the International Geographical Union established a 'Commission on Agricultural Typology'. Under the energetic chairmanship of Jerzy Kostrowicki this Commission sought to formulate a typological framework that would facilitate a standardized classification of world farming systems (Kostrowicki, 1975, 1979). This exercise inevitably generated considerable discussion as to which criteria and which taxonomic methods should be adopted in developing such a framework. In the earlier meetings of the Commission participants presented a diverse array of papers. Some were specifically conceptual, others pondered the problems of identifying key variables and their operational definition, whilst a great many offered regionalisations of home areas using an array of different methods. Amazingly — in view of the complexity of the issues involved and the diversity of viewpoints expressed in responses to two questionnaires and at numerous conferences — by 1971 the first draft of a world typology was produced. After further minor modifications the general structure of the typological model was formally accepted by the Commisssion at its Odessa meeting in 1976. The Commission itself was formally concluded in 1979, but work on applying the model has continued. At the Paris meeting of the International Geographical Union in 1984 Kostrowicki was able to present a map of types of agricultural regions in Europe at a scale 1:2.5 million (Kostrowicki 1982, 1984). Compiled with the help of a number of agricultural geographers, this typological and regional analysis of European agriculture is undoubtedly an impressive achievement — it constitutes a major landmark in the annals of agricultural geography.

The taxonomic system formulated by Kostrowicki is an imposing and elaborate edifice. To date it recognises over 100 different types of agriculture, organized in a hierarchy of three orders (Kostrowicki, 1980), and distinguished according to profiles of scores on a set of 28 attributes. A distinctive feature of the system is that it is an organic and open-ended structure, and allows for the introduction of new types of agriculture as circumstances dictate.

The types currently identified have themselves emerged over a period of time, following extensive field inquiries and detailed analyses of literature and agricultural data from all parts of the world. Implementation of the typological system involves the would-be taxonomist in a straightforward assignment procedure. The types of agriculture recognised by the system are each characterised by a profile of scores on the

28 attributes. Taxonomic units (i.e. farms or farming areas) are identified by seeking out the typal group(s) with which they exhibit the greatest degree of affinity. This is achieved by calculating deviations between the observed and expected profiles. To overcome problems of scale differences and to allow estimates to be made where insufficient data are available, scores on each of the attributes are 'normalised', with world ranges being subdivided into five classes. The matching process is accordingly reduced to the summation of deviations between scores expressed on a 1 to 5 scale. Rather than assigning a particular taxonomic unit to its closest type, the system allows for multiple identifications. This arises because all total deviations that fall below specified thresholds are considered to be significant. These thresholds have been arbitrarily selected and vary for each of the three hierarchical orders of agricultural types recognised within the system. Thus, for first order identifications, deviations of less than 34 are deemed to be meaningful; in the case of the more detailed second and third tiers the equivalent thresholds are reduced to 23 and 12 respectively. Kostrowicki is of the opinion that the six first order types that have been distinguished are not likely to be increased, but that work in less well documented parts of the world is likely to lead to an extension of the lists of second and third order types. The six first order types have been labelled thus: E — Traditional Extensive Agriculture; T — Traditional Intensive Agriculture; M — Market Oriented Agriculture; S — Socialised Agriculture; A — Highly Specialised Livestock Breeding; L — Latifundia. It is to be appreciated that these labels are essentially tags of convenience and do not adequately summarise the essential characteristics of the systems concerned. By way of illustration, the spatial dominance of first order types in Poland is displayed in Figure 2.3 (Kostrowicki, 1982). At present over 20 second order types of agriculture have been distinguished. The number of third order types now exceeds 100. For multiple identifications (i.e. where a taxonomic unit is deemed to be close to more than one type) and for cases where total deviations exceed the designated thresholds, then further processing of the data may be necessary, with a view to isolating transitional or completely new types of agriculture. This latter facility highlights the innate flexibility of the typological scheme, for it ensures that future changes in the nature of farming systems and practices can be accommodated without there being any need to dismantle the whole taxonomic structure. It also follows from this that the typology is particularly well suited to analyses of change and differences in farming structures both in space and time.

The 28 attributes used in the typology were chosen after considerable

Figure 2.3: First Order Agricultural Types, Poland

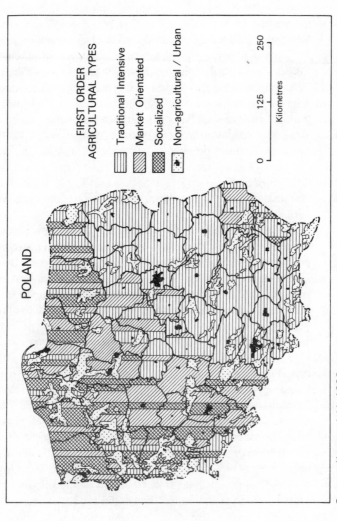

FIRST ORDER
AGRICULTURAL TYPES

Traditional Intensive

Market Orientated

Socialized

Non-agricultural / Urban

POLAND

0   125   250
Kilometres

Source: Kostrowicki, 1983.

debate (Aitchison, 1983) and now form four distinct and balanced clusters — social, operational, production and structural. It is not appropriate here to enter into a detailed appraisal of the terms used to define the various attributes, except for noting that a number of them (e.g. gross agricultural production, livestock units and units of draught animals) are operationalised by reference to standard sets of conversion factors. Levels of gross agricultural production, for instance, are determined by converting all crops, grasses and livestock to grain equivalent units. The four sets of attributes are constituted as follows:

Social Attributes
1. Percentage of total agricultural land held in common.
2. Percentage of total agricultural land in labour and share tenancy.
3. Percentage of total agricultural land in private ownership.
4. Percentage of total agricultural land operated under collective or state management.
5. Number of active workers per agricultural holding.
6. Area of agricultural land per holding (hectares).
7. Gross agricultural production per agricultural holding.

Operational Attributes
8. Number of active agricultural workers per 100 hectares of agricultural land.
9. Number of draught animals per 100 hectares of cultivated land.
10. Number of tractors, harvesters etc. in terms of total horsepower per 100 hectares of cultivated land.
11. Chemical fertilisers: NPK per hectare of cultivated land.
12. Irrigated land as a percentage of total cultivated land.
13. Harvested land as a percentage of all arable land (including fallow).
14. Livestock units per 100 hectares of agricultural land.

Production Attributes
15. Gross agricultural production per hectare of agricultural land.
16. Gross agricultural production per hectare of cultivated land.
17. Gross agricultural production per active agricultural worker.
18. Gross commercial production per agricultural worker.
19. Commercial production as a percentage of gross agricultural production.
20. Commercial production per hectare of agricultural land.
21. Degree of specialisation in commercial production.

Structural Attributes

22. Perennial and semi-perennial crops as a percentage of total agricultural land.
23. Grassland (permanent and temporary) as a percentage of total agricultural land.
24. Food crops as a percentage of total agricultural land.
25. Livestock production as a percentage of gross agricultural production.
26. Commercial livestock production as a percentage of gross commercial production.
27. Gross production of industrial crops as a percentage of total agricultural production.
28. Herbivorous livestock as a percentage of total livestock.

It is necessary to stress that the typological scheme described above has been fine-tuned over a period of years and that as a consequence the majority of published applications are based not on the final version used by Kostrowicki in his analysis of European agriculture, but on slightly different earlier models. Be this as it may, numerous studies have experimented with the scheme, the basic lineaments of which had already been established by the mid-1970s. Notable contributions include appraisals of agricultural types in France (Bonnamour and Gillette, 1980), Belgium (Christians, 1975; Stola, 1983), Poland (Tyszkiewicz, 1975), Bulgaria (Tyszkiewicz, 1979), USSR (Gorbunova, Komleva and Shishkina, 1979), Australia (Scott, 1975, 1983), western regions of the United States (Gregor, 1975), Canada (Troughton, 1975, 1979, 1982), India (Sharma, 1983; Singh, 1979) and Malaysia (Hill, 1982, 1983). These, and many other studies, underline the interest that has been shown in a bold typological venture.

Methods of Classification

Methods of assignment, although arguably lacking in statistical sophistication, have shown themselves to be particularly useful when seeking to derive standard typological frameworks for large and varied sets of taxonomic units. It is for this reason that so many census agencies have adopted assignment strategies in collating agricultural statistics. For comparative investigations and studies of trends or changes through time they are probably more appropriate than methods of *classification* (Figure 2.1). This said, however, it is evident that over the past two decades geographers have made much more use of classificatory procedures in isolating types and regions of agriculture. One reason for this appeal is that in classification the taxonomic units themselves (via their attributes) determine the final group structures — the groups are not imposed as in assignment strategies, they arise from the data. Furthermore, classificatory methods (e.g. cluster analytic and ordination procedures) are especially useful in treating multivariate information. Given the range of attributes that are often included in studies of farming systems this facility is clearly of some significance. Whereas with methods of assignment it is generally necessary to focus on a limited number of attributes and to express these on common scales, with classifications there is much greater flexibility. The data-processing capacity of modern computer systems and the ready availability of software (e.g. packages of classificatory algorithms) have made the actual process of deriving classifications a very simple matter indeed. Carried along by the wave of enthusiasm that accompanied the so-called 'quantitative revolution' a number of agricultural geographers in the 1960s and early 1970s experimented with various types of factor analytic models. As it happens, the potential of such models for dealing with large sets of attributes and for purposes of regionalisation had been demonstrated much earlier by Kendall (1939), Hagood, Danilevsky and Beum (1941) and Hagood (1943). In her study of North American agriculture Hagood (1943) examined patterns of intercorrelation between 104 variables (52 of which were agricultural) and used factor analysis to isolate integrated scales along which the 48 States being investigated could be ordinated. Working with various subsets of variables (in the case of the 52 agricultural variables six subsets were recognised — land use, crops, livestock, tenure, farm values and farm finance) a series of single-factors was derived. Scores on each of these six separately defined factors were subsequently analysed with a view to isolating specific types of agricultural regions. For some unaccountable reason, the analytic strategy pioneered by Hagood failed to attract the attention it deserved, and more than two decades were to pass before factor analytic models were widely adopted by agricultural geographers.

In an early contribution Henshall and King (1966) described patterns of farming on the island of Barbados, using both Q-mode and R-mode forms of factor analysis. The Q-mode solutions (with Varimax rotations) ensured direct classifications (ordinations) of agricultural holdings, whilst R-mode analyses highlighted patterns of intercorrelation between a wide-ranging collection of attributes (Henshall, 1966). The use of factor analytic models in a classificatory context has since become commonplace, with principal components analysis (PCA) in particular being widely employed as both a data-search and a data-reduction procedure (e.g. Aitchison, 1972; Gregor, 1982; Munton and Norris, 1969; Nordgard, 1977; Rey, 1982; Troughton, 1982). Operating inductively and often without explicitly articulated preconceptions, numerous studies have used PCA to seek out underlying dimensions (i.e. components) in multivariate sets of information. These dimensions are not only of interest in their own right — sometimes drawing attention to unforeseen associations — they also constitute scales along which taxonomic units can be arrayed. Classifications of taxonomic units were at first achieved either through visual inspections of bivariate plots of component scores (e.g. Munton, 1972) or by charting the spatial distribution of scores for individual components (e.g. Aitchison, 1972; Troughton, 1975). Similarly, Guermond (1979, 1983) derived a regional typology of agriculture in Normandy by sequentially dichotomising groups of agricultural holdings according to their scores on a set of eight factors. Based on 25 attributes (including land-use, livestock and structural measures), five farm-type categories are distinguished, but with a broad dualism between 'small family farm' holdings and 'large commercial' holdings being particularly highlighted.

Over recent years, agricultural geographers have tended to link PCA models with various cluster analytic procedures. As a data orthonormalisation technique PCA generates scores which can be used to measure Euclidean distances between taxonomic units. These distances serve as coefficients of dissimilarity and can be analysed in a variety of ways to isolate taxonomic clusters. Thus far, the most popular cluster methods have been those that are agglomerative and hierarchical in structure. The process of cluster formation generally involves the gradual fusion of individual units and groups of units, a sequence that is commonly displayed in the form of a dendrogram. Whilst cluster analysis ensures a more rigorous and integrated treatment of component scores, taxonomists have to be prepared to make subjective decisions concerning which algorithm to employ and how many classes or clusters to isolate. Experimenting with several clustering routines (e.g. nearest neighbour, further neighbour, centroid, median, group average and Ward's error sum of squares), Byfuglien and Nordgard (1973) and Aitchison (1975) have shown that these matters can be quite crucial when it comes to identifying and interpreting types and regions of agriculture. Different methods can generate different results, whilst the number of clusters eventually selected will greatly influence assessments of both the patterns and the processes operative within a particular study area. In regard to the choice of methods it would appear that the majority of geographers have elected to base their analyses on 'minimum variance' cluster routines (i.e.

algorithms that seek to minimise within-cluster sums of squares). The extensively implemented 'error sum of squares' procedure developed by Ward (1963) is such an algorithm (Aitchison, 1975; Anderson, 1975; Ilbery, 1981; Norgard, 1977). Other examples of applications of cluster analytic methods include studies by Troughton (1982) on Canadian agriculture, Gillmor (1977) on type-of-farming patterns in Ireland, and Rikkinen (1971) and Talman (1979) on agricultural regions in Finland. It is perhaps worth noting that in classifying agricultural systems some use has also been made of multi-dimensional scaling and discriminant analysis (Anderson, 1975) and of principal coordinates analysis with mixed-mode data (Aitchison and Aubrey, 1982). These procedures are particularly useful where attributes are measured on ordinal or categoric (rather than numeric) scales, and also in assessing the efficacy of typological structures. Analysing the nature and impact of the urbanisation process on farm structures Bryant (1974, 1981) has demonstrated the potential of a divisive cluster analytic procedure, using binary data and an information statistic to derive measures of dissimilarity.

An elaboration which has generated some discussion within the literature concerns the appropriateness or otherwise of introducing a contiguity constraint when seeking to delineate agricultural regions (Byfuglien and Nordgard, 1974). The general consensus would appear to be that a contiguity constraint is particularly undesirable when the aim is to generate or test hypotheses concerning relationships between types of farming and associated factors of production. This is reasonable and the majority of studies have accordingly chosen not to include such a control — which in a hierarchical cluster analysis only allows individual taxonomic units or groups of such units to merge if they have a common border, however small that may be. It should perhaps be emphasised, however, that a contiguity constraint might be useful where, for instance, the purpose of the taxonomic exercise is to delineate agricultural planning regions. Aitchison (1972) also found a contiguity control to be essential in endeavouring to create a more standardised areal frame for a regional analysis of farming systems in Wales. In this case 1027 parishes, highly variable in size and shape, were aggregated to form a more meaningful set of 100 districts. The strategy adopted sought not only to derive compact areal units (hence the contiguity constraint) but also to combine parishes with similar agricultural characteristics.

Whilst it is usual to distinguish between methods of assignment and methods of classification, it should be stressed that the two need not necessarily constitute discrete and unrelated taxonomic strategies. In a study of French agriculture, for instance (SCESS, 1979, 1983), data relating to the percentage dominance of 21 types of agricultural holdings (according to the EEC assignment typology described above) were collated for 712 *Petites Régions Agricoles* (just one of several areal divisions used by the French Ministry of Agriculture for the presentation of statistics). This 712 times 21 matrix was then subjected to a cluster analysis (agglomerative and hierarchical) with a view to isolating broader type-of-farming regions. The structure of the resultant dendrogram suggested that 17 basic types might be recognised. Whilst the results of this analysis were of interest in their own right, the main aim of the whole exercise was not to classify 'small agricultural regions' but to generate a set of categories for use in identifying farming patterns in 3465 cantons. To achieve this, profiles of percentages for individual cantons were matched against equivalent profiles for each of the 17 categories generated in the cluster analysis. Cantons were assigned to particular type-of-farming categories using a Euclidean distance measure. In carrying out this matching process the rather arbitrary decision was taken to exclude cantons with less than ten holdings and to allocate such areas to an ambiguous category *'cantons sans agriculture'*. Figure 2.4 presents a simplified version of a map showing the spatial distribution of the various type-of-farming categories at canton level — for clarity the 18 groups have been aggregated into eight (Rey and Giraudet, 1984). Although this taxonomic strategy is open to a number of criticisms it does serve to illustrate the general point that methods of assignment and classification can be used together within a single investigation. Interestingly, in seeking to isolate three typological orders for the world typology, Kostrowicki and his associates experimented with a number of sophisticated clustering procedures before deciding upon the final hierarchical structure (Bielecka, Paprzycki and Piaseki, 1979; 1980). In this case, the aim was to distinguish first and second order clusters from the larger set of third order types. It will be recalled that these orders serve as *a priori* categories for the identification of types of farms and types of farming areas.

Interpretation and Evaluation Stage

If a general criticism were to be levelled at classificatory studies in agricultural geography, it would be that insufficient attention has been accorded to the question of evaluating the quality of particular typologies or systems of regionalisation. Admittedly, such evaluations can be difficult to make since there are no accepted theoretical or statistical

Figure 2.4: Dominant Farm Enterprise Regions, France

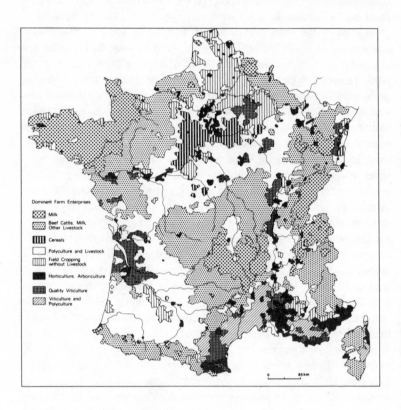

Source: SCESS, 1983; Rey and Girandet, 1984.

standards to which the taxonomist can turn for guidance. The problem is that classifications can never be adjudged 'true' or 'false'. They are structured to satisfy a variety of different purposes and it is only in relation to these purposes that their quality can be evaluated. Various measures might be applied to test the coherence of typal groups (e.g. within and between group dispersion statistics), but in the main, evaluations often have to rest on highly personal judgements as to whether or not elusive epithets such as 'meaningful', 'revealing', 'suggestive' or 'useful' can be attached to particular classifications. Be this as it may, it could be argued that too many agricultural geographers have tended to assume that their typologies or regionalisations are implicitly valid and accordingly are worthy of detailed interpretation. Seldom have efforts been made to justify or defend adopted solutions. All too often studies are based on single analytic strategies, with no attempt being made to test the stability of typal structures using either alternative algorithmic procedures or, in the case of relocation techniques, alternative initial groupings of taxonomic units. Unless such an open and experimental perspective is subscribed to there is the added danger that equally revealing but different (even contradictory) patterns will be overlooked. Increasingly powerful packages of classificatory algorithms, replete with analytic options, have removed many of the computational difficulties that constrained earlier investigation in this regard. Of course, merely subjecting variants of the same basic data set to a range of differing taxonomic methods will not in the end answer the question as to which typological or regional solution is the 'best'. It is at this final evaluative stage that more profound geographical skills and sensitivities find scope for expression and arbitration. Classification is both an art and a science.

References

Aitchison, J.W. (1968) 'The land factor and agricultural production in West Central Lancashire', *Agricultural Geography IGU Symposium*, Research Paper No.5, Department of Geography, University of Liverpool

Aitchison, J.W. (1972) 'The farming systems of Wales' in C. Vanzetti (ed.) (1972) pp. 129-45

Aitchison, J.W. (1975) 'Cluster analysis, regionalization and the agricultural enterprises of Wales' in C. Vanzetti (ed.) (1975) pp. 17-33

Aitchison, J.W. (1979) 'The agricultural landscape of Wales: the structure of agricultural holdings, 1964-74', *Cambria, 6*, 32-53

Aitchison, J.W. (1980) 'The agricultural landscape of Wales: patterns of agricultural production 1964-74' *Cambria, 7*, 1-26

Aitchison, J.W. (1981) 'Triangles, tetrahedra and taxonomy', *Area, 13*, 137-43

Aitchison, J.W. and Aubrey, P. (1982) 'Part-time farming in Wales: a typological study', *Transactions of the Institute of British Geographers, 77*, 88-97

Aitchison, J.W. (1983) 'Model types in world agriculture: problems of definition and case identification', *Geographia Polonica, 46*, 175-86

Anderson, K.E. (1975) 'An agricultural classification of England and Wales', *Tijdschrift voor Economische en Sociale Geographie, 66*, 148-57

Andrianov, B.V. and Cheboksarov, N.N. (1975) 'Historic-ethnographic areas. Problems of historic-ethnographical regionalization', *Sovetskaya Etnografiya, 2*

Andrianov, B.V. (1979) 'African traditional economic-cultural types and the problems of typology of world agriculture', *Geographia Polonica, 40*, 5-9

Baker, O.E. (1926-32) 'Agricultural regions of North America', *Economic Geography, 2-8*

Belshaw, D.G.R. and Jackson, B.G. (1966) 'Type of farming areas: the application of sampling methods', *Transactions of the Institute of British Geographers, 38*, 89-93

Berry, B.J.L. (1962) *Sampling, coding and storing flood plain data*, United States, Department of Agriculture, Farm Economics Division, Agricultural Handbook, p. 237

Bielecka, K., Paprzycki, M and Piasecki, Z. (1979) 'Proposal of new taxonomic methods for agricultural typology', *Geographia Polonica, 40*, 191–200

Bielecka, K., Paprzycki, M. and Piasecki, Z. (1980) 'Applicability of numeric taxonomy methods in agricultural typology: problems, criteria and methods of evaluation' in N. Mohammad (ed.) *Perspectives in Agricultural Geography*, New Dehli

Birch, J.W. (1954) 'Observations on the delimitation of farming type regions: with special reference to the Isle of Man, *Transactions of the Institute of British Geographers, 20*, 141–58

Blaikie, P.N. (1971) 'Spatial organization of agriculture in some north Indian villages', *Transactions of the Institute of British Geographers, 52*, 1–40

Blaut, J.M. (1959) 'Microgeographic sampling: a quantitative approach to regional agricultural geography', *Economic Geography, 35*, 79–88

Board, C. (1970) 'The quantitative mapping of land use patterns with special reference to land use maps: shape analysis, with an application', *Geographia Polonica, 18*, 121–38

Bonnamour, J. (1973) *Géographie Rurale*, Masson et Cie, Paris

Bonnamour, J. Gillette, Ch. and Guermond, Y. (1971) 'Les systèmes régionaux d'exploitation agricole en France: méthode d'analyse typologique', *Etudes rurales, 43–44*, 78–169

Bonnamour, J. and Gillete Ch. (1980) *Les types d'agriculture en France 1970: essai méthodologique*, CNRS, Paris

Boserup, E. (1965) *The Conditions of Agricultural Growth: the Economics of Agrarian Change under Population Pressure*, Allen and Unwin, London

Boserup, E. (1981) *Population and Technological Change: a Study of Longterm Trends*, Chicago University Press, Chicago

Bowman, I. (1932) 'Planning in pioneer settlement', *Annals of the Association of American Geographers, 22*, 93–107

Bryant, C.R. (1974) 'An approach to the problem of urbanization and structural change in agriculture: a case study from the Paris region', *Geografiska Annaler, 56 (Series B)*, 1–27

Bryant, C.R. (1981) 'Agriculture in an urbanizing environment: a case study from the Paris region, 1968–76', *Canadian Geographer, 25*, 27–45

Buchanan, R.O. (1959) 'Some reflections on agricultural geography', *Geography, 44*, 1–13

Byfuglien, J. and Nordgard, A. (1973) 'Region-building: a comparison of methods', Norsk Geografisk Tidsskrift, *27*, 127–51

Byfgulien, J. and Nordgard, A. (1974) 'Types or regions?', Norsk Geografisk Tidsskrift *28*, 157–66

Chisholm, M. (1964) 'Problems in the classification and use of farming-type regions', *Transactions of the Institute of British Geographers, 35*, 91–103

Chisholm, M. (1979) *Rural Settlement and Land Use. An Essay in Location*, Hutchinson, London

Christians, C. (1975) 'La typologie de l'agriculture en Belgique: méthodes, problèmes, résultats' in C. Vanzetti (ed.) (1975) pp. 93–109

Commission of the European Communities (1978) 'Commission decision of 7 April 1978 establishing a Community typology for agricultural holdings', *Official Journal of the European Communities*, 78/463/EEC, No.L148/1–34

Commission of the European Communities (1984) 'Commission Decision of 29 February amending Decision 78/463/EEC establishing a Community typology for agricultural holdings', *Official Journal of the European Communities*, 84/260/EEC, No.L128/1–35

Coppock, J.T. (1960) 'The parish as a geographical-statistical unit', *Tijdschrift voor Economische en Sociale Geografie, 51*, 317–26

Coppock, J.T. (1964a) *An Agricultural Atlas for England and Wales*, Faber, London

Coppock, J.T. (1964b) 'Crop, livestock and enterprise combinations in England and Wales', *Economic Geography, 32*, 65–81

Cruickshank, J.G. and Armstrong, W.J. (1971) 'Soil and agricultural land classification in County Londonderry', *Transactions of the Institute of British Geographers, 53,* 179-94

Dennett, M.D., Elston, J. and Speed, C.B. (1981) 'Climate and cropping systems in West Africa', *Geoforum, 12,* 193-202

Duckham, A.N. and Masefield, G.B. (1970) *Farming Systems of the World,* Chatto and Windus, London

Fletcher, A.A. (1983) *Agricultural information flows in North Wales,* Unpublished PhD thesis, University of Wales

Gasson, R. (1973) 'Goals and values of farmers', *Journal of Agricultural Economics, 24,* 521-42

Gillmor, D.A. (1977) *Agriculture in the Republic of Ireland,* Akademiai Kiado, Budapest

Gorbunova, L.I., Komleva, M.V. and Shishkina, L.V. (1979) 'Agricultural typology of the USSR', *Geographia Polonica, 40,* 83-93

Gregor, H.F. (1970) *Geography of agriculture: themes in research,* Prentice-Hall, Englewood Cliffs, New Jersey

Gregor, H.F. (1975) 'A typology of agriculture in western United States in world perspective' in C. Vanzetti (ed.) (1975) pp. 173-185

Gregor, H.F. (1982) *Industrialization of United States agriculture,* Westview, Boulder, Colorado

Grigg, D.B. (1969) 'The agricultural regions of the world: review and reflections', *Economic Geography, 45,* 95-132

Grigg, D.B. (1970) *The Harsh Lands: a study in agricultural development,* Macmillan, London

Grigg, D.B. (1974) *Agricultural Systems of the World: an evolutionary approach,* Cambridge University Press, Cambridge

Grigg, D.B. (1984) *An Introduction to Agricultural Geography,* Hutchinson, London

Guermond, Y. and Massias, J.P. (1973) 'L'utilisation du sol en France: comparaison de deux méthodes de traitement de l'information', *L'Espace Géographique, 2,* 267-73

Guermond, Y. (1979) *Le système de différenciation spatiale en agriculture, la France de l'Ouest de 1950 a 1975,* Paris

Guermond, Y. (1983) 'Les types d'expolitations agricoles en Normandie', *Geographia Polonica, 46,* 133-47

Haggett, P. (1963) 'Regional and local components in land-use sampling: a case study from the Brazilian Triangulo', *Erdkunde, 17,* 108-14

Hagood, M.J. Danilevsky, N. and Beum, C.O. (1941) 'An examination of the use of factor analysis in the problem of subregional delineation', *Rural Sociology, 6,* 216-33

Hagood, M.J. (1943) 'Statistical methods for delineation of regions applied to data on agriculture and population', *Social Forces, 21,* 288-97

Hart, J.F. (1975) *The Look of the Land,* Prentice-Hall, Englewood-Cliffs, New Jersey

Harvey, D. (1968) 'Pattern, process and the scale problem in geographical research', *Transactions of the Institute of British Geographers, 45,* 71-81

Harvey, D. (1969) *Explanation in Geography,* Edward Arnold, London

Henshall, J.D. (1966) 'The demographic factor in the structure of agriculture in Barbados', *Transactions of the Institute of British Geographers, 38,* 183-95

Henshall, J.D. and King, L.J. (1966) 'Some structural characteristics of peasant agriculture in Barbados', *Economic Geography, 42,* 74-84

Hidore, J. (1963) 'The relationship between cash grain farming and landforms', *Economic Geography, 39,* 84-95

Hill, R.D. (1982) *Agriculture in the Malaysian region,* Akademiai Kiado, Budapest

Hill, R.D. (1983) 'The Malaysian region and the world typology of agriculture', *Geographica Polonica, 46,* 21-47

Ilbery, B.W. (1981) 'Dorset agriculture; a classification of regional types', *Transactions of the Institute of British Geographers, 6,* 214-7

Ilbery, B. (1983a) 'Goals and values of hop farmers', *Transactions of the Institute of*

British Geographers, 8, 329–41

Ilbery, B. (1983b) 'A behavioural analysis of hop farming in Hereford and Worcestershire', *Geoforum*, 14, 447–59

Ingram, D.R. (1984) 'Simplifying ternary diagrams', *Area*, 16, 175–80

Jackson, B.G., Barnard, C.S. and Sturrock, F.G. (1963) *The pattern of farming in the eastern counties: a report on a classification of farms in eastern England*, Farm Economics Branch, School of Agriculture, Cambridge

Jenkins, T.N. (1982) *The classification of agricultural holdings in Wales*, University College of Wales, Department of Agricultural Economics, Aberystwyth

Kendall, M.G. (1939) 'The geographical distribution of crop productivity in England', *Journal of the Royal Statistical Society*, 102, 21–48

Kostrowicki, J. (1975) 'The typology of world agriculture; principles, methods and model types' in C. Vanzetti (ed.) (1975) pp. 429–79

Kostrowicki, J. (1979) 'Twelve years activity of the IGU Commission on Agricultural Typology', *Geographia Polonica*, 40, 235–53

Kostrowicki, J. (1980) 'A hierarchy of world types of agriculture', *Geographia Polonica*, 43, 125–62

Kostrowicki, J. (1982) 'The types of agriculture map of Europe', *Geographia Polonica*, 48, 79–91

Kostrowicki, J. (1984) 'Types of agriculture in Europe; a preliminary outline', *Geographia Polonica*, 50, 132–49

McCarty, H.H. (1954) 'Agricultural Geography' in P.E. James and C.F. Jones (eds.) *American Geography: Inventory and Prospect*, Syracuse University Press, Syracruse, 258–277

Morgan, W.B. and Munton, R.J.C. (1971) *Agricultural Geography*, Methuen, London

Morgan, W.B. (1977) *Agriculture in the Third World: a spatial analysis*, Bell and Sons, London

Munton, R.J.C. and Norris, J.M. (1969) 'Analysis of farm organization: an approach to the classification of agricultural land in Britain', *Geografiska Annaler*, 52, 95–103

Munton, R.J.C. (1972) 'Farm systems classification: a multivariate analysis' in C. Vanzetti (ed.) (1972) pp. 89–106

Nordgard, A. (1977) 'Types and regions of Norwegian agriculture', *Norsk Geografisk Tidsskrift*, 31, 15–26

Perpillou, A. (1952) 'Essai d'etablissement d'une carte de l'utilisation du sol en France', *Acta Geographica*, 18, 110–15

Perpillou, A. (1970) *Carte de l'utilisation agricole du sol en France, seconde moitié du XXe siècle*, CNRS, Paris

Perpillou, A. (1977) *Carte de l'utilisation agricole du sol en France, seconde moitié du XXe siècle*, CNRS, Paris

Reeds, L.G. (1964) 'Agricultural geography: progress and prospects', *Canadian Geographer* 2, 51–63

Rey, V. (1982) *Besoin de terre des agriculteurs*, Economica, Paris

Rey, V. and Giraudet, E. (1984) 'Complexité des systèmes agricoles' in *140 cartes sur la France rurale*, Géo-media, Paris

Rikkinen, K. (1971) 'Typology of farms in central Finland', *Fennia*, 106, 5–44

Ruthenberg, H. (1971) *Farming Systems in the Tropics*, Clarendon Press, Oxford

SCESS. (1979) 'La classification des exploitations agricoles selon leurs orientations technico-économiques', *Cahier de Statistique Agricole*, No. 146

SCESS. (1983) 'La carte "systèmes de production agricole"', *Cahier de Statistique Agricole*, 3, 17–20

Scott, P. (1957) 'Agricultural regions of Tasmania', *Economic Geography*, 33, 109–21

Scott, P. (1975) 'The application of world agricultural typology to Australia' in C. Vanzetti (ed.) (1975) pp. 297–309

Scott, P. (1983) 'The typology of Australian agriculture', *Geographia Polonica*, 46, 7–19

Sharma, B.L. (1983) 'A typological analysis of agriculture in the Rajasthan State', *Geographia Polonica, 46*, 71–7

Singh, V.R. (1979) 'Agricultural typology of India', *Geographica Polonica, 40*, 113–31

Spencer, J.E. and Stewart, N.R. (1973) 'The nature of agricultural systems', *Annals of the Association of American Geographers, 63*, 529–43

Stola, W. (1983) 'Essai d'application des méthodes typologiques à l'étude comparée sur le développement des agricultures belge et polonaise', *Geographia Polonica, 46*, 159–73

Talman, P. (1979) 'Areal livestock combinations on Finnish farms', *Fennia, 157*, 155–70

Tarrant, J.R. (1974) *Agricultural Geography*, David and Charles, Newton Abbot.

Taylor, J.A. (1952) 'The relation of crop distributions to the drift pattern in south-west Lancashire', *Transactions of the Institute of British Geographers, 18*, 77–91

Thomas, D. (1963) *Agriculture in Wales during the Napoleonic wars: a study in the geographical interpretation of historical sources*, University of Wales Press, Cardiff

Troughton, M.J. (1975) 'Approaches to an agricultural typology for Canada' in C. Vanzetti (ed.) (1975) pp. 357–84

Troughton, M.J. (1979) 'Application of the revised schemes for the typology of world agriculture to Canada', *Geographica Polonica, 40*, 95–111

Troughton, M.J. (1982) *Canadian Agriculture*, Akademiai Kiado, Budapest

Tyszkiewicz, W. (1975) 'Types of agriculture in Poland as a sample of the typology of world agriculture' in C. Vanzetti (ed.) (1975) pp. 391–408

Tyszkiewicz, W. (1979) 'Agricultural typology of the Thracian Basin: Bulgaria as a case of the typology of world agriculture', *Geographia Polonica, 40*, 171–90

Van Hecke, E. (1983) 'The structure of Belgian agricultural production: a cartographic representation', *Tijdschrift voor Economische en Sociale Geografie, 74*, 213–6

Vanzetti, C. (1972) *Agricultural Typology and Land Utilization*, Centre of Agricultural Geography, Verona

Vanzetti, C. (1975) *Agricultural Typology and Land Utilization*, Centre of Agricultural Geography, Verona

Varjo, U. (1977) *Finnish Farming: Typology and Economics*, Akademiai Kiado, Budapest

Varjo, U. (1984) 'Changes in farming in Finland in 1969–1975', *Fennia, 162*, 103–15

Ward, J.H. (1963) 'Hierarchical grouping to optimize an objective function', *Journal of the American Statistical Society, 58*, 276–46

Weaver, J.C. (1954a) 'Crop combination regions in the Middle West', *Geographical Review, 44*, 175–200

Weaver, J.C. (1954b) 'Crop combination regions for 1919 and 1929 in the Middle West', *Geographical Review, 44*, 560–72

Weaver, J.C. (1956) 'The county as a spatial average in agricultural geography', *Geographical Review, 46*, 536–65

Whittlesey, D. (1936) 'Major agricultural regions of the earth', *Annals of the Association of American Geographers, 26*, 199–240

Wood, W.F. (1955) 'The use of stratified random samples in a land use study', *Annals of the Association of American Geographers, 45*, 350–67

3 DIFFUSION OF AGRICULTURAL INNOVATIONS

G. Clark

The popular image of farming is one of tradition and conservatism with the continuity of life more noticeable than the changes. Today this image is a myth since farming in the Third World as much as in the developed countries is embracing new methods at a steady pace. Indeed, in some cases new ideas bring their own problems of over-production of food and environmental pollution. New methods and practices are a very important part of agriculture since they are involved in establishing the type and systems of farming practised, they have a key role to play in structural changes, and governments frequently deem it expedient to have policies to speed the adoption of new methods. The world food problem, it is claimed, could be solved by applying better farming methods in the Third World. The present chapter is concerned to study the causes and effects of new methods of farming. It will not study the invention of new methods but will deal with their progressive adoption and then their replacement by still more modern techniques.

It could be argued that the study by geographers of the adoption of innovative methods in farming has been too tightly focused so far (Agnew, 1979; Blaikie, 1978). This narrowing of perspective needs to be reversed in four critical areas if the subject is to be comprehensively analysed. First, the definition of an innovation should be widened. Second, the methods used to analyse the topic need to be reassessed. Third, the relationship between innovations and economic development must be thought out more clearly. Finally, the topic could be more firmly linked to economic and social life as experienced by individual people. This chapter will examine each of these four areas in turn describing the current style of analysis and indicating where progress is needed.

Definitions

An innovation is usually defined as any new idea, organisation, practice or piece of equipment which is used to achieve some social, economic or cultural objective. In practice the majority of studies of innovations have been of 'things' rather than 'practices'. This is understandable since objects such as new types of machinery are more easily identified

70

on the ground than, say, a new way of sowing wheat, and machinery is more often recorded in censuses and company records. It is usually easier to study the equipment for farming than a method of farming.

This statement is least true, however, when we consider the earliest applications of machinery. For example, who invented the mechanical reaper? Such questions raise two serious problems. The orthodox answer may depend on which country you are in. There is some pride involved in being seen as a nation of inventors and so rival claims to the title of the 'real' inventor of the reaper or mechanical milker are advanced in different countries. This nationalistic slant derives partly from simple patriotism but also from a genuine confusion of definition. What constitutes a steam engine or combine harvester? How far is it necessary to hop before you have built the first aeroplane? The early days of any venture will be marked by equipment of limited capacity and formidable unreliability. As in photography or television, different systems may be used to achieve the same end. Therefore, it is difficult to pin-point the starting point from which diffusion is to be measured since there are often several different definitions of the equipment. These uncertainties are greater still when studying new breeds of animal or new methods of organisation since there are no clear reference points or performance standards against which to test whether an organisation is really a co-operative or a herd is actually pure-bred Hereford cattle. The difficulties are compounded if the innovation started some time ago when the historical record which has survived is incomplete or is set out in an ill-defined form or using words which are difficult to interpret unambiguously.

The importance of matters of definitions is partly a function of the way studies of diffusion have evolved. A frequent topic for study has been the earliest adopters of innovations. This group of pioneers and entrepreneurs has exerted a powerful fascination for geographers and other academics. This may be an amalgam of admiration for their daring and a practical interest in identifying this group so that future innovations can be more speedily marketed.

One way of studying diffusion has been to study the number of people who have adopted as a proportion of some total population. Many studies have suggested that the growing acceptance over time of an innovation can be modelled by an S-shaped curve as in Figure 3.1. A slow early growth at the left-hand end of the diagram (the area with the definitional problem) is followed by a steeply rising curve of rapid adoption and finally a tailing off at the right-hand end of the diagram as the last few people catch on.

Figure 3.1: Cumulative Adoption of an Innovation

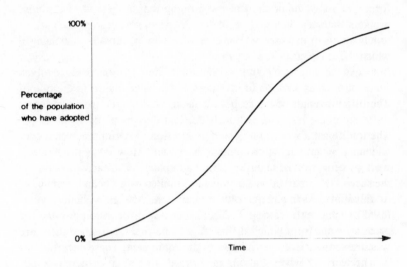

It is frequently claimed that this S-shaped curve is a logistic curve which is the classic form of the growth of an epidemic (Cassetti, 1969). However, this seems unlikely since a logistic curve requires that the 'disease' (in our case, the innovation) be equally potent throughout its life and that all peoples are equally susceptible to catching it (or adopting it, in our case). These are unlikely assumptions so it is better to describe Figure 3.1 as an S-shaped curve and leave its exact mathematical form for further study. Such diagrams, which are common in the diffusion literature, assume that the total population through which the innovation will spread can be defined and is constant throughout the period of adoption. This is inherently unlikely since some people, to whom the innovation is relevant at the start, will change business or will no longer be able to afford it. New circumstances will allow other people to join the group to whom the innovation applies. Similarly, the innovation itself may change. It may be improved in terms of its robustness, reliability or capacity, or its price may be reduced. These changes will expand or contract the size of the population which might conceivably buy it. However, most studies of innovations have assumed a static population through which the innovation occurs. This question of an innovation's population needs more attention in the future.

This section has argued that there are important areas of definition which need more attention in diffusion studies. This is particularly critical in the very early days of an innovation's development when its exact form is rapidly evolving as different solutions are sought to a technical problem by rival scientists, perhaps in several countries. We also need to have a clearer idea of just what is meant by the population through which the innovation spreads.

Traditional Approaches

The traditional approaches to studying innovation diffusion have been widening, particularly during the last ten years. However, the earliest work by geographers was strongly cartographic in nature. Maps showing the successive distributions of the innovation were studied for evidence of a spatial pattern to adoption (Hägerstrand, 1952). Work by Sauer (1952) sought to trace the spread of agricultural developments in the prehistoric and early-modern periods. The spatial dimension of diffusion was stressed as the distinctively geographical contribution to such studies.

The work of Hägerstrand (1953) took the study a stage further by inquiring into two matters. First, he considered whether there was a common spatial pattern to the diffusion of different innovations. Second, he investigated how the diffusion took place and hence why the distinctive spatial pattern emerged. Both questions continue to exercise geographers' minds.

The first issue is the extent to which inductive generalisations can be made about spatial patterns of adoption. This heralded a move towards analysing diffusion as a recurrent process rather than just a series of incidents or case studies. Hägerstrand saw the principal spatial pattern of diffusion as a contagious process — what he called the neighbourhood effect. An innovation was accepted by those nearest the people who had already adopted it. Hence the probability of adopting an innovation declined as distance from a previous adopter increased. Hägerstrand also devised a statistical method of simulating such a spreading pattern — the Monte Carlo technique. This used random numbers to simulate the diffusion. Later workers, such as Morrill (1970), reverted to Hägerstrand's earlier metaphor of a wave of adoption. The analogy is with the wave spreading over the surface of a pond after a stone has been thrown in. The point of impact of the stone is the equivalent of the first person to adopt. The timing of adoption is therefore a function of a person's distance from the first innovator. It soon became clear, however, that this contagious pattern of diffusion was a very inaccurate description of the spread of certain types of innovation and that other patterns could be detected.

The most common of these patterns came to be called a hierarchical pattern (Berry, 1972). This described diffusion that started in the most important place at the top end of a hierarchy of areas and then worked its way down to the less important places. Griliches (1957) showed how a new variety of hybrid maize was first made available in the principal maize-growing areas and only later in those regions which were less suited to the crop. Similarly, many developments in transport and telecommunications have been introduced first to major conurbations and then have spread down the urban hierarchy to progressively smaller settlements. Pedersen (1970) combined these two spatial patterns into a single formula:

$$I_{ij} = k \cdot P_i \cdot P_j^{\alpha}/d_{ij}^{\beta}$$

where I_{ij} = the interpersonal interaction between zones i and j
P_i and P_j = the populations of zones i and j
d_{ij} = the distance between zones i and j
α and β = exponents

As α tends to zero so interaction is a function of distance, that is, a contagious pattern of diffusion. As β tends to zero, so interaction becomes a function of the population of the two places, that is a hierarchical pattern. Any particular diffusion, he argued, might contain elements of both contagious and hierarchical diffusion. Actually measuring the variables and exponents could prove difficult but this gravity-model formulation marked a step forward in our thinking about the spatial structure of diffusion. Pedersen's studies of diffusion in Chile and other Latin American countries and the studies of Sheppard (1976) and Webber and Joseph (1977) suggest some generalisations about hierarchical and contagious spread. The former seems to be dominant at the upper end of the urban hierarchy, in the more highly developed countries, and in the more recent historical periods. Contagious diffusion is more important lower down the urban hierarchy (i.e. from village to neighbouring village), in earlier historical periods and in less developed countries.

Hägerstrand's other major contribution to analysing diffusion was to consider the mechanism which generated spatial patterns. He modelled the process as a social one based on the primacy of word-to-mouth communication. He was studying small-scale factors in inter-war Sweden and he assumed that, since migration and mobility were limited, face-to-face contact would be the critical means of passing information about the innovation. The city-based mass media either had not been invented in this period or were in their infancy. The central role of personal contact in furthering acceptance of the innovation fitted in well with a contagious (i.e. neighbour-to-neighbour) pattern of adoption.

Later workers in the field have argued that economic factors as well as social ones will affect diffusion and should also be considered. The difficulty with this statement is that it sets up an opposition between the economic and the social which is both artificial and rather difficult to measure. Sheppard's (1976) attempt to separate the two used population size as one of the surrogate measures of both the potential profitability of the innovation and the social structure of settlements. The definition of an 'economic factor' also presents difficulties for particular case studies. Does this refer to the capital cost of an innovation or its running costs? Does it refer to the short-term or long-term profitability of the innovation? Is the judgement to be based on money alone or on workload, for example, and is the innovation to be compared only with other agricultural investments or with any other possible investment? There are several possible definitions of the economic factor in any decision to invest in an innovation and each could lead to different motives and diffusion patterns.

Hägerstrand placed great stress on 'communication' as the key mechanism in diffusion. This has presented researchers with practical difficulties in testing the effectiveness of information flows which are still very poorly understood. Communications between neighbours, with salesmen and within families may take place and yet leave no record for the researcher to find. Hägerstrand's face-to-face conversations are the most difficult to monitor and record. The immense volume of material from television and radio is sometimes not available after it has been broadcast; the written word is easier to study in newspapers, leaflets and advertisements. It is, however, true that the information flows which impinge on farmers are increasing in type and volume. Local and national newspapers and radio compete today with television viewdata services (e.g. Prestel), specialist farming magazines and advertising literature. The common feature of all these is how they have become increasingly urban in character. The major providers of information are based in the cities and particularly in the principal city of a country. Here you find the headquarters of broadcasting agencies, telecommunications organisations and international publishing companies. Multinational companies selling machinery or agrochemicals are increasingly based in the larger cities and the new technologies of cable and satellite television seem to be evolving in an equally centralised way. This centralisation of control and information implies that the quality of information available to farmers is becoming increasingly homogeneous over large areas of the earth, in contrast to former times when much less information was available and even this spread slowly and unevenly.

Hägerstrand's stress on face-to-face communication also raises the question of to whom farmers talk and, more importantly, to whom they listen. It is reasonable to suggest that there was once a time when small-scale farmers with little formal education and limited mobility did talk principally to neighbours whose views were potentially relevant to their situation since they too farmed on the same small scale, under the same environmental conditions and with comparable technology and markets. Today we can suggest that farmers are a much more diverse group of people. Some operate on a large scale and some on a smaller scale; some are young and well-educated while others are older or more in the mould of practical men; many farmers are specialists in a few crops or in livestock rather than all being mixed farmers as previously. One's neighbours are less likely to be the type of farmer whose concerns and experiences are similar or relevant to one's own. Such farmers are likely to be those who have the same size and type of farm as oneself irrespective of where they are located relative to each other (Maclennan, 1973). Conferences and specialist shows provide fora for these people to meet and compare notes.

If we argue that some information and conversations are more influential than others then our attention is focused less on the information itself than on how it is received and what use is made of it in the decision on whether or not to adopt the innovation. Jones (1967) has suggested that the decision process can be modelled as a series of stages (Figure 3.2).

Early in the process, the mass media may well be the most important in alerting a farmer to a possible solution to a problem. At the trial and evaluation stages more specialist information might be useful. At any point in this progression, the innovation might fail to meet the farmer's perceived need and so be rejected. Only if it passes all these tests would it be adopted.

This model is perhaps too neat since it implies that the process is wholly rational with the decision maker expending effort on evaluating the innovation in proportion to its attractiveness. At each stage rational criteria are apparently used to judge the innovation's worth. Although there is still much to be learnt about how people make decisions, it is rarely as logical a process as Jones suggests. The criteria people use are varied (Gasson, 1973). Sometimes a farmer will opt for a development

Figure 3.2: Stages in the Decision to Adopt

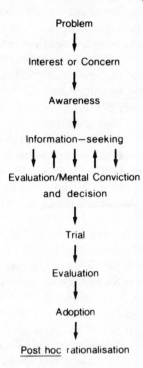

Source: Adapted from Jones, 1967.

which will raise his output (Harle, 1974), at other times one which
reduces his costs of production or his workload will be favoured. It is
also clear, as Audley (1971) has argued, that what makes up a mind is
its history, and evidence too — but only if that is in the right direction.
Or to put it another way, minds quite often come already made up.

Audley was dealing with decision-making by individuals and that is
still the correct scale of analysis for most farming, but there is a trend,
particularly in the USA, for farming to be conducted by companies of
considerable size rather than by individual farmers (Smith, 1980).
Decision-making within organisations is more complex and is usually
analysed in terms of how those at the top of the organisation deal with,
and react to, the stream of information and recommendations flowing
up to them from the diverse interest groups (e.g. marketing, finance and
production) which make up the farming corporation (Downs, 1971). The
application to diffusion studies of the models produced by organisation

theory might also be productive.

Both this analysis and that of individual decision-making are based on the assumption that time is available to assess the innovation and the farmer is a free agent to determine investment in his own best interests. This would, of course, be as misleading a view of the individual farmer's situation as it would be of the farming corporation. Time is never plentiful; thorough searching for information can be protracted and tends to be skimped. The weather or marketing conditions may force themselves on the farmer. Decisions therefore tend to be taken before deadlines imposed by others and hence to be sub-optimal (Flowerdew, 1976). The decision on whether to adopt an innovation should be viewed as the best that can be done in the time available and with the information to hand. It will reflect the priorities and experiences of the individual farmer rather than what is objectively in his best long- or short-term interests.

The geographer studying past decisions to adopt innovations is then faced with another practical problem, *post hoc* rationalisation. This is the tendency for people to justify a decision once taken, to re-interpret why they took it, and so present the researcher with a false picture of the decision. This hinders the researcher in his task of reconstructing how the innovation appeared to those trying to decide whether to accept it.

This idea of the importance of how an innovation appears to people moves our attention to the innovation itself. Some are small-scale affairs, others involve a major change in procedures. Some are cheap and some are expensive. Rogers (1962) has suggested that five aspects of any innovation are important in explaining how fast it will diffuse. Rapid acceptance, he argued, would be associated with the most profitable innovations which are compatible with other aspects of the farm, easy to understand and describe, and which are divisible. The importance of the communicability of an innovation's qualities and the ease with which these can be grasped is obvious. The divisibility of an innovation refers to the ability to test it on a small scale before a major commitment is made. If the new crop can be tried on a small plot or the machine leased for a month, then the innovation should diffuse more rapidly, if its qualities are sound. A new method of husbandry or a computerised accounting system, however, may be less susceptible to small-scale trial and so its adoption will therefore be slower. The importance of the profitability of an innovation is also worth stressing but with some caution. Profitability can be measured in various ways as has been described already and not all innovations will rate equally well by each of these criteria.

Figure 3.3: Cumulative Diffusion as a Compound Process

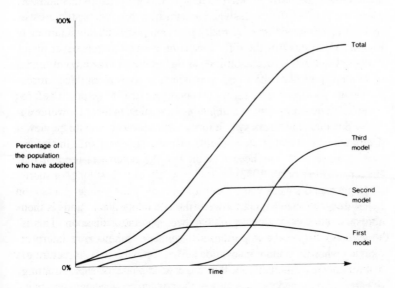

Whichever method is used in practice, what is important is how an innovation is expected to perform over time. This element of forecasting is essential for a proper analysis of diffusion but has been largely ignored in previous studies. Exceptions are work by Jones (1962), Chapman (1974) and Clark (1977). These studies showed that forms of innovation which minimised the possibility of loss or failure in farming were highly sought after. Farmers were willing to forgo immediate profit, capital growth and lighter workload in order to avoid losses. Farming is still a risky business and memories of depression, bankruptcy and adverse weather conditions are never far from the minds of farmers. The effect of the uncertainties these thoughts induce in farmers should be more fully examined in future research.

A discussion of the properties of the innovation itself should not leave the impression that these are fixed: an innovation will often evolve over time. Early models are typically slow, cumbersome, unreliable, limited in performance and expensive. Over the years, refinements are made as competition between manufacturers stimulates improvements and early customers feed back suggestions for enhancements. Therefore, the objective qualities of any innovation, its profitability for example, will change over time; whether its qualities as perceived by potential innovators will similarly improve is less clear. One often finds that some users of an early model will switch to later ones. Consequently, the diffusion of the innovation may really be a compound phenomenon as in Figure 3.3. Instead of the diffusion being a single process (the upper curve in Figure 3.3) it is better seen as the related but distinct diffusions of several models of the equipment which in total form the complete diffusion. In time even the improved models of the product will be superceded by a new way of doing things and so an innovation will eventually decline as its users switch to the next innovation. This period of decline has been called the paracme of the innovation and, in the few studies made of it, has been modelled as the reverse of the adoption (Baker, 1977; Ilbery, 1982).

Hägerstrand's work was concerned to explain the timing of adoption by reference to the distance between the potential adopter and previous adopters. Parallel to this theme has been a socio-economic approach which has stressed the importance of people's 'innovativeness' in explaining when they adopt innovations. This approach starts from the idea that for any innovation there will be a few early pioneers, a larger number of early adopters and then the bulk of the population, leaving only a small number of laggards who will adopt much later (Figure 3.4). The cumulative growth of users will therefore take the form shown in Figure 3.3.

This partitioning of the population into groups with greater or lesser innovativeness has formed the basis for detailed descriptions of the personal characteristics of the groups. Generally the early innovators and later adopters were seen as complete opposites (Rogers, 1962). Early adopters were younger, better educated, wealthier, more modern in outlook, more cosmopolitan, more specialised in their business and more open to professional and technical sources of information. These personal profiles do present some problems however. Although they succeed in correlating personal traits with early innovation, it is less clear what is the causal link between them. There is a danger of circular arguments sometimes and the late adopters are sometimes dealt with too harshly. The distinctly pejorative term 'laggards' is used often to describe late adopters when in fact these people may include those for whom the innovation is only marginally suited in terms of its cost and capabilities (Yapa and Mayfield, 1978).

Of course, both this socio-economic approach and that of Hägerstrand assume that the innovation is simultaneously made available to the entire

Figure 3.4: Types of Adopter Defined by Time of Adoption

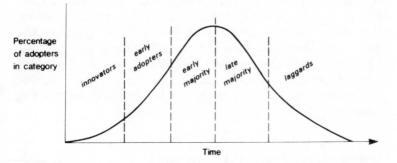

population. There are many cases where this is not so and consequently demand-based approaches are inadequate. The innovation may be supplied to some areas before others as part of a conscious marketing strategy by a propagating agency. An explanation for the diffusion of such innovations must start from the propagator's sales policy. The priorities and perceptions of those who manage the sales strategy need to be examined.

L. Brown (1981) has proposed that the critical factor in analysing the supply of innovations is whether the organisation concerned has a centralised structure or a decentralised one. Brown suggests that within a centralised system the marketing strategy will be determined by a set of constraints and one or more marketing goals. Constraints might include limited financial resources and hence a need either to market only in the home area or to use an existing sales force. The marketing goals would determine the order of entry to such market areas as are financially feasible. Areas might be entered in order of their profitability, their density of potential sales or their volume of potential sales. Prestige might require the marketing of a new variety of wheat to start in the heartland of wheat growing where sales would be dense and plentiful though competition might reduce profit margins.

A decentralised system of control (e.g. a franchise system for marketing a service like grain drying or machinery maintenance) would lead to a more random pattern of supply. There would be a less strong adherence to a grand strategy than in a centralised system. Much would depend on which areas produced entrepreneurs to take on the franchises. This converts diffusion into a two-stage process; first, the acceptance of the franchise (instead of other activities) by entrepreneurs and, second, the acceptance of the innovation by the public. L. Brown (1981) also identified another kind of supply organisation where profit is not a motive. Official farm advisory services would fall into this category. In theory the pattern of supply in such cases could be either random or uniform but in practice such organisations tend to have limited resources and a strong professional commitment to maximising the service they give. This leads often to a pattern of supply based on the need for the service, which often resolves itself into a hierarchical distribution of the service not unlike the marketing strategies used by centralised profit-seeking organisations.

Innovations and Economic Development

The previous section was concerned with the adoption of innovations and what affected the demand for them and their supply. Attention was focused on making inductive generalisations about individuals' actions and then explaining these. This scale of analysis is the one most commonly used in the geographical literature on innovation and, for that reason, has become rather a straitjacket (M. Brown, 1981). In this section it is proposed that the geographer's interest in innovations should extend beyond adoption at the level of individuals and encompass also the broad-scale consequences of innovations for economic development. There have been many theories of agricultural development and several have accorded innovation a key role in development; some of these will be reviewed here.

Boserup's (1965) work was based on her experience of peasant societies in Africa and Asia and she stressed how innovations were used to cope with growing population density. As population rose in a peasant society more people had to be fed from a given agricultural area. Starvation was kept at bay by a series of technical improvements which raised the productivity of the farmland. These took the form of converting pasture to arable, reducing or eliminating the fallow period, growing more productive crops and where possible fertilising and irrigating farmland (Miracle, 1967). In this model, population growth encourages the use of new husbandary in order to raise food supply. Malthus's model, in contrast, saw agricultural innovation as a process whose slowness acted to check the otherwise rapid growth of population. Boserup's model has not gone uncriticised and it is clearly specific to a particular kind of farming economy. It is based on a number of assumptions about the independence of population growth and food supply, the preference of peasants for leisure rather than profit, and their knowledge of more productive or labour-intensive methods which they choose not to use unless necessary. Boserup's model only applies in a pre-commercial subsistence agriculture where other methods of dealing with the population-food problem are not feasible or are inadequate (e.g. specialisation, contraception, migration to towns or breaking in new farmland). The model is likely to be most applicable in historical studies although the work of Chayanov in the 1920s suggested that similar attitudes to work and profit were found in Russia at least as late as pre-1914 (Grigg, 1982).

In contrast to Boserup's model, Myrdal's (1957) theory of development is advanced by its proponents as being of contemporary relevance and it has been modified recently in analyses of the Green Revolution in the Third World (Frankel, 1971; Griffin, 1974; Yapa, 1977). Myrdal's model is summarised in Figure 3.5.

Development is regarded as a self-reinforcing process since those who develop first gain benefits which allow them quicker access to the next round of innovations — 'to him that hath shall be given'. In the context of the Green Revolution, Griffin noted the factor and social bias to many of those farming innovations which comprised the Green Revolution. Innovations provide most benefits to those who control capital and land (landlords, merchants and larger-scale farmers) rather than those who control labour (smaller-scale farmers and labourers). L. Brown used the term 'adoption rent' to express the idea that the early adopters of innovations such as high-yielding cereals gain a double benefit. First, they raise their output and, second, they gain a greater profit per unit output. Later adopters will gain less benefit as the increase in food supply will have become sufficient to lower market prices. The early adopters will therefore gain windfall profits and so will be able to afford later innovations. Griffin and Yapa extended this idea further by noting how the Green Revolution required not only new varieties of seed but also considerable capital to provide the pesticides, irrigation and fertilisers necessary to reap the potential of these varieties. Financial reserves or access to credit were needed, the latter being most easily and cheaply available to those more progressive and credit-worthy farmers. Again a social bias can be seen to operate with previous expansion based on the adoption of one set of innovations providing conditions favouring further expansion through a later set of innovations. The social effect of this may be to polarise the farming community into larger, wealthier farmers and poorer ones, whereas previously a more equal social status among farmers was evident. This social division may be reinforced since the expanding

Figure 3.5: Myrdal's Process of Cumulative Causation

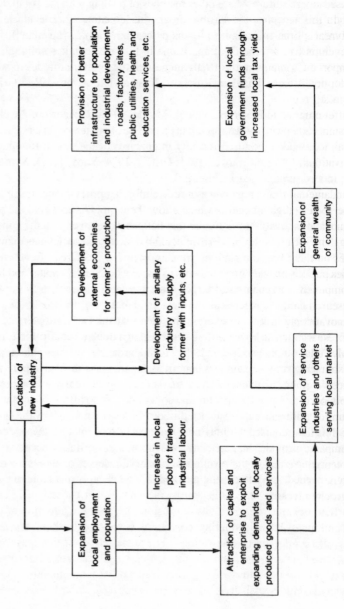

Source: D. Keeble, *Models in Geography*, Methuen, London, 1967, p. 258.

farmers may need to employ workers for their extra land, in which case a capitalist sector of agriculture will emerge based on wage labour and selling food. Other farms will remain more clearly subsistence and family-based operations.

In this model of the Green Revolution the cumulative character of Myrdal's general theory is worked out using innovations as the key to development. There is clearly a paradox here in that innovations are an important component of development, but their differential and biased adoption creates those inter-personal inequalities which are one symptom of regional and personal underdevelopment. Furthermore, the differential adoption is not seen as a matter of chance, far less of mere distance or personal entrepreneurship. The structure of society, it is claimed, makes it easier for some groups than others to adopt innovations.

Indeed, the role of innovations in agricultural society is even more fundamental than has been suggested above. Innovations not only alter the society they spread through, but that society controls the kinds of innovations that are developed and affects the rate at which they are taken up. The influence over the kinds of innovations that are developed centres on the control of funding for agricultural research. In most countries publicly funded research into new methods of farming is an extensive component of agriculture's hidden subsidy. In theory, public funds for research could be directed either into profit-making innovations or into innovations which are neutral or will help small-scale farmers. The case of research in the USA into 'alternative' agriculture is instructive here (McCalla, 1978; Paarlberg, 1978; Youngberg, 1978). When the actual disbursement of money is analysed, it appears that the projects which receive grants are those which a pressure group supports, where public opinion is favourable (or, at best, not hostile) and where a broad spectrum of political and commercial support can be generated. The kinds of developments that are fostered are usually those which commercial companies support because profits can be foreseen from their sale. Sales potential is equated with usefulness. Developments in husbandry which favour small, mixed farms or which dispense with machinery or agrochemicals are less often funded.

Innovations which cost money have to be repaid before any change in enterprise can be considered. A bulk milk tank and mechanised milking allow much larger dairy farms but they also lock the farmer into dairying. He is unlikely to be able to throw away this dairy investment next year if the price of milk falls. In so far as innovations allow specialisation, they also increase the risk of farming — the mixed nature of farms used to be their insurance policy against one crop failing. There is an insistent need, therefore, for a system of price stabilisation operated by government. Another paradox therefore emerges. Innovations may be adopted in order to reduce certain risks and dependencies inherent in primitive farming — crop failure, hence higher yielding varieties; poor harvesting weather, hence faster harvesting machinery. Yet these same innovations reduce a farm's ability to react to price signals from the market and so they force a greater dependence on government support to reduce price shocks and on scientific research to keep one step ahead of both other farmers and the resistance of pests to insecticides. The consequences of scientific research or government support failing are more severe than ever. Galbraith (1974) has developed similar ideas in the context of major industrial companies and these are relevant to the structural consequences of innovations on farming. He notes how technological development (frequently supported by government grants and advisory agencies) favours larger businesses (for reasons discussed above) and also demands planning and freedom from risk. This implies that, as the farmer comes to control increasing amounts of technology, he will spend less of his time doing manual work and more time planning. He will also seek ways to minimise risk by controlling his input and output prices. Structural changes such as co-operatives, vertical integration, contract farming, hedging on futures' markets and share farming will be favoured. Each strategy acts either to reduce risk, or to shift some of the farmer's risk to another party (e.g. a food manufacturing company).

Innovations and Individuals

The previous section argued that our understanding of innovations is incomplete if the two-way relationship between technological developments and the structure of society and the economy is ignored. Similarly, there is a need to study the effect of innovations on the life of the individuals. When adoption is dealt with at the level of inductive generalisation, one inevitably loses detail, and, what is more important, understanding. There is much to be gained by analysing separately factors such as distance, innovativeness and supply organisations in order to see how each affects diffusion and adoption. However, for the individual farmer these factors may not exist (innovativeness and information are synthetic concepts created by academics) and the real-world phenomena from which they derive (newspapers for example) do not exist in discrete compartments. The factors interact in different ways for different people and the resulting complexity and diversity is an essential feature of agricultural change.

A good example of this complexity is provided by Gallaher's (1961) study of 'Plainville', the pseudonym for a farming community in Missouri in 1954–5. He describes the efforts of an official extension agency to introduce new methods into an area of many small and rather backward family farms. His first perspective focuses on the supply-side.

> The improved pasture program, for example, was introduced to Plainville farmers as a 'permanent pasture program', with little or no effort to clarify the key term 'permanent'. It was assumed that farmers knew the program would provide fall and winter pasture which, combined with spring and summer pasture, would make grazing lands 'permanently' available throughout the year. This assumption proved erroneous. Several leading farmers interpreted 'permanent' to mean perennial grasses which could provide serviceable grazing twelve months of the year, but because of their skepticism rejected the innovation from the beginning.
>
> Others, who made the same assumption but were less skeptical, were well involved before learning that the program provided only fall and winter grass. Some of them reacted negatively, inferring that the innovation had been deliberately misrepresented. Much confusion was eliminated by changing the program title to 'improved pasture', but only after a negative base had been established . . .
>
> Despite the early semantic confusion and high cost, the improved pasture program was on its way to general acceptance when the drought struck in 1952 . . . Because of the drought participants, some with sizable investments, did not realize their expectations. . . .
>
> Thwarted by natural events, frustrated by loss of money some farmers turned their hostility against the main innovative agency, Extension. A few early acceptors insist that Extension oversold them; others say that Extension is not as aware of local conditions, weather and otherwise, as it should be before making recommendations; and some criticize the agents:
>
>> Them fellers got to git good cooperation on them programs to make their reports to the higher-ups look good. . . .
>
> Scientific practices introduced by government agencies convey to Plainvillers man's potential control over the land, and in doing so give them new respect for it. There is, however, always the weather,

and when Plainvillers ponder man's relationship to this element the present blurs against the past. This bothers farmers ('No matter what you do, you got to have the right weather for it'), and makes them question the gamble of cash outlays for fertilizers and other expensive programs if weather controls are lacking.

The questions raised here concern the perception not only of the innovation — it may be misunderstood and fail — but also of the propagating agency and the general farming background. Gallaher focuses on the central importance of the farmers' perception of the dire consequences of the variability of the climate for their husbandry. This view (perhaps it is more an obsession) has a major effect in influencing the farmers' attitudes to new farming practices. He stresses the part played by obstinacy, ignorance and surprise in the farmers' behaviour and in how they (mis-)interpreted the motives of the extension agency. Gallaher also investigated the farmers' perspective on the same process and in particular he studied their reasons for adopting new machinery.

Thus, men who had a few years ago gained prestige by working long hours at hard physical labor are remembered today as 'slaves' to hard work and long hours, and if one of them still manifests these qualities he is ridiculed as 'behind the times'. Industriousness is now measured by the speed and ease of labor performance, which is achieved by an implement complex built around the tractor. . . .

The Plainviller's desire to own machinery is further supported by a strong emphasis upon individualism. Individualism is so highly valued that close relatives may duplicate major and expensive equipment. Farmers, in fact, agree unanimously that 'it's best for every man to own his own equipment'. . . .

Finally, it is obvious that some Plainville farmers want new and expensive machines for prestige reasons. Equipment is an important symbol of achievement in the competitive agricultural system, and its possession is a major criterion in the evaluation of individual success. As a measure of social worth its importance stems from the fact that most people are secretive regarding their incomes. Since incomes are a topic for speculation and gossip among neighbours, the amount is inferred by reckoning from expenditures. Inasmuch as a significant part of the farmer's budget is allotted to machinery and its upkeep, judgments of neighbor upon neighbor invariably stress the quantity, size, and expense of equipment, all known variables.

This study stresses how complex and 'irrational' decision-making really is since factors such as avoiding being seen as behind the times, independence and social prestige are very important.

These extracts provide a counterweight to the supply and demand approaches and the high-level theoretical contributions described earlier in this chapter. They stress the complexity of innovation at the level of the individual in distinction to the inevitably simplifying picture presented by macro-level analyses. Both approaches are essential for fully describing how innovations spread.

Future Research

It will be clear that innovation diffusion is an important field in terms of geographical theory, public policy and its assessment, and also personal welfare. It is hoped that research into innovation diffusion will continue and three key areas for further study can be identified.

There is a need for better understanding of the processes that influence the kinds of innovation made available or marketed to potential adopters. If the character of a new product will affect its geographical impact (e.g. which groups or areas benefit most), then the processes which determine that character are worthy of study. This implies research into the allocation of public funds for research, the political and scientific control over research programmes and the way research agendas are established in public and private organisations. One could also study the origins of public policies (e.g. subsidies and specialist advisors) for the promotion of certain kinds of innovations but not others.

The second area requiring study is the effect of (particularly technical) innovation on the spatial structure of economies and societies and the reverse influence of spatial structure on innovation diffusion. The ideas of Galbraith (1974) are useful here in establishing some hypotheses which could be tested. He suggests that increasing technological sophistication requires certain changes in the economy, namely more planning, a greater stability in production and prices, and a more general freedom from risk. Also there are less predictable consequences from mechanisation such as greater dependence on other people (technicians, scientists and politicians, for example) and a slowness to react to externally generated changes in the economic environment. These hypotheses are *prima facie* applicable to the agricultural sector and their elaboration and testing would be welcome.

Finally, there is a need for a fuller appreciation of how decisions on marketing and adopting innovations are taken by organisations and individuals. Such studies in a humanist tradition seem to lack the rigour, objectivity and quantifiability of statistical sources which allow for generalisations to sets of cases. Yet what they lack when judged by the requirements of inductive argument and statistical hypothesis testing, they make up for through their insights into real decision-making.

Innovation diffusion is an important component of historical, economic and cultural geography that invites a healthy inter-disciplinary approach involving geographers, economists, anthropologists and sociologists. It needs to be studied at a wide variety of scales and many quantitative, theoretical and practical techniques. It is likely to remain a key area within agricultural geography.

Acknowledgements

I should like to acknowledge the assistance of Mrs Anne Jackson (Cartography Unit, Department of Geography) who drew the diagrams, and Mrs Jean Burford and Miss Maxine Young who typed this paper.

References

Agnew, J.A. (1979) 'Instrumentalism, realism and research on the diffusion of innovations', *Professional Geographer, 31,* 364–70
Audley, R.J. (1971) 'What makes up a mind?' in F.G. Castles, D.J. Murray and D.C. Potter, *Decisions, Organisations and Society,* Penguin, Harmondsworth, Ch.5
Barker, D. (1977) 'The paracme of innovations'. *Area, 9,* 259–65
Berry, B.J.L. (1972) 'Hierarchical diffusion: the basis of developmental filtering and spread in a system of growth centres' in N.M. Hansen (ed.), *Growth Centres in Regional Economic Development,* The Free Press, New York
Blaikie, P. (1978) 'The theory of the spatial diffusion of innovations — a spacious cul-de-sac', *Progress in Human Geography, 2,* 268–95
Boserup, E. (1965) *The Conditions of Agricultural Growth: the economics of agrarian change under population pressure,* Allen and Unwin, London
Brown, L.A. (1981) *Innovation Diffusion - a New Perspective,* Methuen, London
Brown, M.A. (1981) 'Behavioural approaches to the geographic study of innovation diffusion: problems and prospects' in K.R. Cox and R.G. Golledge (eds.), *Behavioural Problems in Geography,* Methuen, London, Ch.6, pp. 123–44
Cassetti, E. (1969) 'Why do diffusion processes conform to logistic trends?' *Geographical Analysis, 1,* 10–105
Chapman, G.P. (1974) 'Perception and regulation: a case study of farmers in Bihar', *Transactions of the Institute of British Geographers, 62,* 71–94
Clark, G. (1977) *The amalgamation of agricultural holdings in Scotland, 1968-1973,* Unpublished PhD thesis, University of Edinburgh
Downs, A. (1971) 'Decision making in bureaucracy' in F.G. Castles, D.J. Murray and

D.C. Potter, *Decisions, Organisations and Society,* Penguin, Harmodsworth

Flowerdew, R.T.N. (1976) 'Search strategies and stopping rules in residential mobility', *Transactions of the Institute of British Geographers, 1,* 47–57

Frankel, R.F. (1971) *India's Green Revolution,* Princeton University Press, Princeton, New Jersey

Galbraith, J.K. (1974) *The New Industrial State,* Penguin, Harmondsworth.

Gallaher, A. (1961) *Plainville Fifteen Years Later,* Columbia University Press, New York

Gasson, R. (1973) 'Goals and values of farmers', *Journal of Agricultural Economics, 24,* 521–37

Griffin, K. (1974) *The Political Economy of Agrarian Change; an essay on the Green Revolution,* Harvard University Press, Cambridge, Mass.

Grigg, D. (1982) *The Dynamics of Agricultural Change,* Hutchinson, London

Griliches, Z. (1957) 'Hybrid corn: an exploration in the economics of technological change', *Econometrica, 25,* 501–22

Hägerstrand, T. (1952) *The Propagation of Innovation Waves,* Lund series in Geography, Series B. No.4

Hägerstrand, T. (1953 in Swedish, 1967 in English) *Innovation Diffusion as a Spatial Process* (Translated by A. Pred and G. Haag) University of Chicago Press, Chicago

Harle, J.T. (1974) 'Further towards a more dynamic approach to farm planning — a technically based model of the farm firm', *Journal of Agricultural Economics, 25,* 153–63

Ilbery, B. (1982) 'Hop growing in Hereford and Worcestershire', *Area, 14,* 203–11

Jones, G.E. (1962) *Bulk milk handling,* University of Nottingham, Department of Agricultural Economics, Farm Report No. 146

Jones, G.E. (1967) *The adoption and diffusion of agricultural practices,* World Agricultural Economics and Rural Sociology Abstracts, Reading, 1–34

Macalla, A.F. (1978) 'Politics of the agricultural research establishment' in D.F. Hadwiger and W.P. Browne (eds.) *The New Politics of Food,* Lexington Books, Lexington, Mass., Ch.7, pp. 77–91

Maclennan, D. (1973) 'A re-examination and reconsideration of the neighbourhood effect', *Geographical Journal, 139,* 583–5

Miracle, M. (1967) *Agriculture in the Congo Basin — tradition and change in African rural economies,* University of Wisconsin Press, Madison

Morrill, R.L. (1970) 'The shape of diffusion in space and time', *Economic Geography, 46,* 259–68

Myrdal, G. (1957) *Economic Theory and Underdeveloped Regions,* Methuen, London

Paarlberg, D. (1978) 'A new agenda for agriculture' in D.F. Hadwiger and W.P. Browne (eds.), *The New Politics of Food,* Lexington Books, Lexington, Mass., Ch.12, pp. 135–40

Pederson, P.O. (1970) 'Diffusion within and between National Urban Systems', *Geographical Analysis, 2,* 203–54

Rogers, E.M. (1962) *The Diffusion of Innovations,* The Free Press, New York

Sauer, C.O. (1952) *Agricultural Origins and Dispersals,* American Geographical Society, New York

Sheppard, E.S. (1976) 'On the diffusion of shopping centre construction in Canada', *Canadian Geographer, 20,* 187–98

Smith, E.G. (1980) 'America's richest farms and ranches', *Annals of the Association of American Geographers, 70,* 528–41

Webber, M.J. and Joseph, A.E. (1977) 'On the separation of market size and information availability in empirical studies of diffusion processes', *Geographical Analysis, 9,* 403–9

Yapa, L.S. (1977) 'The Green Revolution: a diffusion model', *Annals of the Association of American Geographers, 67,* 350–9

Yapa, L.S. and Mayfield, R.C. (1978) 'Non-adoption of innovations: evidence from discriminant analysis', *Economic Geography, 54,* 145–56

Youngberg, G. (1978) Alternative agriculturalist: ideology politics and prospects in D.F. Hadwiger and W.P. Browne (eds.), *The New Politics of Food*, Lexington Books, Lexington, Mass., Ch.20, pp. 227–46

4 FARMING SYSTEMS IN THE MODERN WORLD

M.J. Troughton

There is no doubt as to the continuing significance of agriculture throughout the modern world. Agriculture remains the dominant form of employment and farming the basis for a 'way-of-life' for perhaps two-thirds of the world's households. Because of their basic importance, attention is focused upon the majority, traditional farming systems and on how they might be improved, with the primary objective being increased production and generation of food surpluses. On the other hand, there have long existed in the economically developed world farming systems which have achieved those basic goals, which have shifted from the 'way-of-life' for the majority to the 'way-of-earning-a-living' for the few, and on behalf of a dominantly non-farm population. To a large extent, these *farming systems in the modern world,* in one form or another, provide models for the transformation of traditional agriculture.

However, while attention is being paid to the appropriate choices for, and on the process of, the transformation of traditional systems, modern agriculture itself, is undergoing a set of equally fundamental, and generally even more rapid, changes. It is this latter transformation affecting agriculture in the modern, i.e., developed, world that is the subject of this chapter, the justification being that, although the process affects only a minority of the world's farmers and farmland, it is profoundly altering the model(s) towards which a much larger segment may be moving, and reflects the disproportionate economic and political influence of the developed world.

Consequently, the discussion is of changes affecting modern farming systems found primarily in developed countries. Overall, these changes are characterised as *'Agricultural Industrialisation'.* Simply defined, this refers to the process, currently active in most developed countries, whereby agriculture (farming) is transformed from an activity generally carried out at a small scale and moderate to low level of capital intensity, to one in which the major proportion of production comes from a reduced number of large-scale highly capitalised units. Although this 'definition' suggests a straightforward or evolutionary shift in farm operation, even those who emphasise its operational nature, see 'industrialisation' as not just any, but *the* change affecting modern agriculture. The stronger viewpoint (advanced here) is that industrialisation involves a fundamental reorientation, in that an industrial model, which previously functioned alongside, rather than as the basis for, modern farming, is now being adopted as an integral part of farm production, as well as for activity beyond the farm gate. The consequences go beyond the individual farm operation, to affect the overall structure of agricultural systems, leading to fundamental change in both the physical and socio-economic rural environment, sufficient to warrant the term 'revolution' when assessing the nature and impacts of these changes (Gregor, 1982).

In an attempt to justify such a strong statement, the chapter addresses the concept and the reality of agricultural industrialisation according to the following sequence: (1) two models which attempt to describe the chronology and processes of the so-called 'agricultural industrial revolution' and its widespread impacts, and (2) selective exemplification of industrialising systems according to a preliminary typology.

Three Revolutions

Although agriculture, especially in its broadest sense as the essence of distinctive rural systems, is often depicted as the product of slow, incremental or evolutionary changes, the role of radical change is well-established. The literature provides comprehensive support for both a 'First' and a 'Second Agricultural Revolution', the term justified by evidence of radical changes affecting substantial portions of mankind, based on successive adoption of new forms of agricultural activity. The thesis advanced here is that agricultural industrialisation has the characteristics of a 'Third Agricultural Revolution'.

Our first model (Figure 4.1) postulates the existence and seeks to describe the essential characteristics of the three 'revolutionary' stages of agricultural development; i.e., (1) the *Beginnings of Agriculture* more than 10,000 years ago, its adoption and subsequent diffusion over much of the earth; (2) the radical shift from *Subsistence* to *Market oriented* production, spreading from eighteenth century England to dominate areas of European settlement by the mid-twentieth century; and (3) *Industrialisation*, affecting both capitalist and socialist developed agricultural economies in the post-Second World War period. Justification of the term 'revolution' to describe contemporary change involves a brief review of previous revolutions and the identification of what constituted(s) the radical changes associated with each, in turn.

The fundamental significance of the First Agricultural Revolution is acknowledged in that, following the innovation and successful adoption

Figure 4.1: Three Revolutions

	1. BEGINNINGS & SPREAD	2. SUBSISTENCE TO MARKET	3. INDUSTRIALISATION
Time:	*Pre 10,000 BP to 20th C.*	*c. 1650 AD to present.*	*1928 to present.*
Key Periods:	*Neolithic. Medieval Europe.*	*18th C. England. 19-20th C. in "european" settlement areas.*	*Present day.*
Key Areas:	*Europe. South & East Asia.*	*Western Europe and North America.*	*U.S.S.R. & Eastern Europe. North America & Western Europe.*
Major Goal:	*Domestic food supply and survival.*	*Surplus production & financial return.*	*Lower unit cost of production.*
Characteristics:	*Initial selection & domestication of key species.*	*Critical improvements, merchantilistic outlook, & food demands of Industrial Revolution, replace subsistence with market orientation.*	*Collective (socialist) & corporate (capitalist) ideologies & common agro-technology favour integration of agricultural production into total food-industry system.*
	Farming replaces hunting & gathering as way-of-life & basis of rural settlement and society.	*Agriculture part of sectoral division of labour: individual family farm becomes "ideal" for way-of-life & for getting a living.*	*Emphasis on productivity & production for profit, replace agrarian structure & farm way-of-life.*
	Agrarian societies proliferate, & support population growth.	*Commercial agriculture develops growing reliance on technological inputs & infrastructure.*	*Collective/corporate production utilises economies of scale, capital intensity, labour substitution & specialised production on fewer, larger units.*
	Subsistence agriculture: labour intensive, low technology, communal tenure.		

of farming methods in key centres or 'hearths', agricultural activity diffused to replace hunting and gathering in most parts of the world. This shift has been causally related to the growth of human population in relation to the increased carrying capacity of agricultural production. It led to widespread sedentary rural settlement and a dominantly agrarian way-of-life, albeit providing the base for a small non-farming urban population (Grigg, 1974; Harlan, 1975). The norm was subsistence agriculture, which although exhibiting considerable spatial variation, was (is) characterised by the direct application of human labour, a low level of technology, and an emphasis on communal tenure. Interestingly, those traditional systems with the highest levels of productivity, notably the so-called 'hydraulic systems' of intensive, irrigated crop regimes (Wittfogel, 1956), although developing sophisticated modes of organisation of land and labour, proved among the most resistant to changes stemming from a 'Second Agricultural Revolution'.

Despite a degree of ecological stability, traditional agriculture exhibited critical limitations: its inherently low yields and an inability to meet the demands of the increasingly mercantilistic, individually inclined society that emerged in Europe in the late Middle Ages. The urge to realise the monetary value of land, both as capital and return to labour, induced shifts in tenure and farm operation, which, in turn, required a market orientation to provide the basis for the disposal of surplus production (Jones, 1967). The Second Agricultural Revolution, beginning in the seventeenth century, became rapidly and intimately associated with the Industrial Revolution. The shift from subsistence to market production went hand-in-hand with the demand for labour and adequate food supply for a rapidly growing industrial workforce. Nevertheless, the two revolutions had distinct underpinnings and characteristics which led to parallel, rather than fully integrated development.

Independent innovations within farming led to increases in production and encouraged the shift from communal-peasant to individual commercial farm operations. These included the introduction of new crops (roots, legumes) and their use in crop rotations which improved soil structure, fertility and output. In turn, livestock, wintered on the fodder crops, could be kept in larger numbers and improved. Other land improvements included drainage, marling and manuring (Fussell, 1965; Jones, 1967). While initial commercial production tended to be concentrated on larger estates, it soon became obvious that the potential lay in broad reorganisation of peasant holdings into individual farms. The latter could provide the base for improvements and increased output upon which market surplus and capital generation was based. In some cases the transition involved dispossession of former tenants and their replacement by a new entrepreneurial class of farmers, e.g. in southern England (Cobbett, 1830). Elsewhere, an emancipated peasantry realised their own commercial opportunities, e.g. in Denmark (Jensen, 1937). Underlying the shift was the demand for food from a rapidly increasing urban, industrial labour force, which outstripped the agricultural capacity of Western Europe and stimulated expansion of market-oriented agriculture in newly colonising areas, especially North America and Australasia, where agricultural settlement was based increasingly on the ideal of the individual family operation (Troughton, 1982).

Commercial farming developed side by side with urban-industrial society and delivery of agricultural output to market and consumer was increasingly facilitated by the application of industrial technology. This included direct inputs to production, notably manufactured farm implements, but particularly the improved infrastructure of transportation, storage, processing and distribution of output. Nevertheless, throughout the formative period of commercial agriculture in the second half of the nineteenth century and its maturing in the first half of the twentieth, strong operational and ideological distinctions were maintained between agriculture and industry. The latter, after an initial 'cottage' stage, became dominantly large-scale, based on large inputs of labour, fossil-fuel and machines in specific factory locations. In contrast, farming, despite a market orientation, actually proliferated through the establishment of greater numbers of relatively small, family units, using small amounts of labour and capital, resulting in modest production per farm unit. Dispersion and a 'cottage scale', rather than concentration, represented major distinctions in the nature of the firm between agriculture and industrial production (McCarty and Lindberg, 1966), and the individual family farm became the integral social, economic and political unit in the structure of both new and modified agrarian societies, i.e., in North America, Australasia and much of Western Europe.

By the twentieth century industry, both primary and manufacturing, was marked by new levels of spatial concentration, product specialisation, and widespread adoption of the integrated assembly line. Still in contrast, farming, despite continuous adoption of mechanical implements, including the tractor and inorganic fertilisers, attained its greatest efficiencies through their application on relatively small-scale mixed operations (e.g., in the US Mid-west and Denmark) (Higbee, 1958). Although industrial activity was applied to some output systems, for example, prairie wheat utilising elevator storage, long distance rail and lake freighter transportation, and centralised milling, most agricultural processing remained local and small-scale prior to the Second World War.

As significant as the operational differences between farm and factory was the growth and maintenance of a belief in the separate nature and values of rural-agricultural versus the urban-industrial sectors of capitalist society. Although this so-called 'myth of the family farm' has recently been challenged and its economic and political supports were already being weakened in the 1930s (Vogeler, 1981), political emancipation involving the transformation from peasant to freehold farmer in Europe and pioneering on individual land grants in North America fostered a strong attachment to both the concept and reality of a separate class of independent farmers and an associated rural society serving the farm community and sharing its values. In a number of countries, for example in Scandinavia, this class consciousness gave rise to farmer-based political parties (Jensen, 1937) and elsewhere to powerful rural movements (e.g., The Grange in the USA). Rural values were (and still are) widely espoused and received political allegiance even in societies shifting rapidly to a predominantly urban condition.

Postulation of a 'Third Agricultural Revolution' based on the industrialisation of farming involves explicit recognition that processes currently under way embody another fundamental shift, including radical changes to the model of capitalist commercial agriculture described above. A number of factors are of critical importance including technology, its applications, and changing economic parameters, but above all the substitution of an industrial for an agrarian model of organisation for farming and its integration into a total agro-food system, which also involves a shift in the ideological context in which farming operates.

Significantly, the industrial model was initiated where individual commercial agriculture was only weakly developed and in an overtly political context, i.e., in the USSR where agriculture was collectivised in 1928. Although some land in Russia was redistributed following the Bolshevik Revolution, collectivisation was from a former landlord-peasant communal system to state ownership and was applied in an ideological context, based on the socialist belief in the elimination of private ownership and an attempt to establish a rural proletariat (Lydolph, 1979; Symons, 1972). Of greatest long-term importance, however, was the meshing of the political rationale with the industrial model of economic organisation. The collective farm was the rural equivalent of the urban factory in terms of its scale, the organisation of its labour force and its role within an integrated, planned system of production and consumption. For a time political concerns, notably the elimination of peasant attitudes, dominated, but the period since the Second World War has seen the strengthening of the economic objectives and the emergence of the consistent socialist viewpoint, i.e., that agriculture constitutes a unified sector in which input supply, production (the farm level) and agricultural processing should be planned as an integrated industry. This model denies any inherent distinctions between industry components and supports the large-scale, state or collective farms as the only 'logical' production units. Collectivisation represents the most consistent version of agricultural industrialisation, in which political and economic ideologies complement one another, and which, theoretically, minimises distinctions between rural and urban society. It has been consolidated in the USSR and was extended to most countries of socialist Eastern Europe between 1945 and 1955 (Committee for World Atlas of Agriculture, 1966).

Collective industrialisation, however, affects only a minority of modern agricultural systems. Industrialisation has also developed in a capitalist framework, albeit with greater problems of rationalisation in both political and economic terms. The capitalist equivalent of the large integrated state enterprise is the large corporation. Corporate farming, in which the large-scale business enterprise is applied to farming, appeared in the 1920s in the United States, to some extent as a logical successor to the colonial and neo-colonial plantation (Gregor, 1974) particularly in California as the basis for horticulture production, using large numbers of hired labourers, often seasonal or migrants, and seeking to achieve the economies of scale on huge holdings, through use of mechanical inputs and linkages to national processing and/or distribution networks. This corporate farming was antithetical to the family farm and its community and social impacts were strongly criticised (Goldschmidt, 1947). Nevertheless, it continued to receive explicit and implicit support from government, industry and agricultural research in the USA and corporate involvement, including that from companies based primarily outside farming, grew in the 1940s and 1950s. However, although there are still a significant number of very large and integrated corporate enterprises, associated particularly with specialised horticultural and livestock production (Gregor, 1974; Smith, 1980), corporate farming *per se* has not expanded to become the dominant industrial form of capitalist farm organisation either within or outside the USA.

Rather, a more general set of factors involving the combined effects of the application of agricultural technology and the changing economics of agricultural production have come into play and provide the basis of more widespread and pervasive change; specifically the capital intensification of production by individual farm operators (Gregor, 1982). A number of factors associated with the Second World War, including the pressure for more output and shortages of farm labour, stimulated rapid application of mechanisation and inputs of fertilisers to farming and increased scale in the agribusiness sectors. The result was a rapid increase in productive capacity, such that it came to exceed both domestic and global commercial demand. The situation was apparent in the USA and Canada in the early 1950s and has since become general in 'western' developed regions. This technologically induced state of overproduction led to a fundamental change in the economics of commercial agriculture (Johnson, 1972). Formerly, increased production inputs were the means to achieve greater output and meet the demands of an expanding market in which all reasonably efficient producers, large or small, could share. Now, with surplus capacity, increased output meant lower market prices. The solution offered to the individual was to lower costs by means of more efficient production based on secondary (capital) inputs. Unfortunately, the real problem was (is) an inelastic demand situation, in which survival rests on an ability not only to compete in terms of lower output costs but to control a larger share of the market. The situation has rapidly evolved whereby attempts to be more efficient in terms of lower unit cost of production have involved even greater individual energy-intensive inputs, as well as product specialisation. This has fuelled a situation which required fewer farms and farmers, with those who survive being those best able to adapt to input efficiencies and economies of scale. The means to this end results in the adoption of elements of an industrial model by the individual farm operator. High capitalisation, low margins and a high debt load have become widespread characteristics. While rationalisation at the farm level has proceeded rapidly, it has been far exceeded in the agribusiness sectors which have experienced very rapid consolidation, often to an oligopoly with, consequently, a propensity to control large segments of the overall agricultural system, including the volume and flow of farm production (Crittenden, 1981; Hightower, 1975; Mitchell, 1975).

The result has been that, although political rhetoric and sentiment continue to support the individual, 'independent' family farm production unit, economic and political reality favours a continuously reduced set of producers operating in an industrial mode and in an economic framework in which market control rests with a few large units, largely outside the farm sector (Gregor, 1981; Vogeler, 1981). This situation represents a sharp divergence from the 'traditional' basis of capitalist agriculture, and the speed with which the change has occurred is also sufficient to warrant the term revolutionary. One result is a convergence between the modes of agriculture found in both modern capitalist and socialist systems, and a major threat to the continued existence of distinctive agrarian societies.

Process and Response in Agricultural Industrialisation

The characteristics advanced to support the case for agricultural industrialisation as a third, revolutionary stage in the development of modern agricultural systems (Figure 4.1) are identified more precisely in a model which depicts the specific processes and which also summarises some impacts of industrialisation (Figure 4.2).

The starting place is a purely utilitarian viewpoint of modern agriculture as an input-output system or 'assembly-line' sequence, in which secondary inputs from the agricultural supply sector (machines, fertiliser, capital) are applied at the production level (the farm) to produce food output, processed and distributed to meet the demands of a predominantly urban society, according to the conditions of either the capitalist market economy or the socialist redistributive system (Figure

Figure 4.2: Model Framework for Industrialising Agriculture in Developed Countries

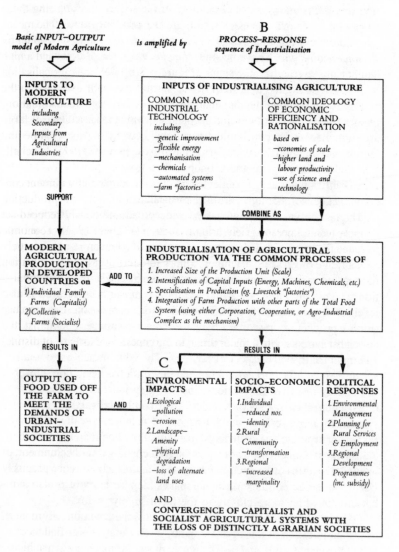

4.2(a)). In this context which, it is suggested, is the dominant bureaucratic view of the nature and rationale of agricultural activity, industrialisation represents an overt 'amplification' of the sequence. Amplification takes place through processes which are the combined outcome of an *Agro-Industrial Technology* and an *Ideology of Economic Efficiency and Rationalisation* which have become *common* attributes of both capitalist and socialist agricultural systems (Figure 4.2(b)).

It might be argued that the post-war period has been marked by an emphasis on so-called 'fundamental differences' between the ideologies of socialist and capitalist developed countries, which in agricultural terms are embodied as 'irreconcilable distinctions' between state-planned collective and individual private farm enterprises. However, an examination of the technological and economic directions employed, reveals a remarkably similar set of elements and criteria adopted and a common convergence towards agricultural industrialisation.

The last 25 years have witnessed widespread and concerted efforts to apply technology to agricultural production (Higbee, 1963). This has included the extension of elements already part of commercial agriculture, i.e., a range of versatile machinery, inorganic fertilisers and the results of crop and animal science. In each area the range and level of sophistication have grown; more powerful machines, fertilisers generating higher levels of crop response, more efficient livestock conversion. But despite surplus production, research and development continues. Government and agribusiness control the process and compete in efforts to improve and innovate in every aspect of farming. The commitment is to totally scientific (and efficient) agriculture and the result has been more powerful, energy-intensive innovations, including an increasingly varied set of agricultural chemicals (pesticides, growth enhancers, etc.), and new systems for large-scale animal husbandry. The latter, so-called 'factory farms' incorporate automated systems that allow one man to 'care for' thousands of animals. In that respect, agro-technology has become post-industrial in terms of its electronic base and minimal labour requirements. Use of an increased range of inputs makes the farmer more productive but also ties him more closely to the input-supply sector.

Of particular significance to our argument is the common nature and overall application of this agro-technology. This may be verified by the fact of the manufacture and use of a range of similar machines, chemicals and related technology in all developed countries, and the adoption of specific items and even complete systems without regard to ideological boundaries (Dohrs, 1982). It is paradoxical that although many key technological advances have been 'western' in origin, they have sometimes found their most obvious application and unequivocal acceptance by collective farm managers, operating at the factory scale, geared to integrated production, and with no concern for the impact on smaller, family-farm competitors. In capitalist countries, nevertheless, despite the obvious threat of large scale inputs and organisation to many existing farms, farmers as a group tend to respond positively to the call for increased levels of technology and the status it imparts to its (successful) users (Kramer, 1980).

Agro-technological development has not occurred in a vacuum; those engaged in its development and promotion see widespread application of science and technology as part of, and thoroughly consistent with, an ideology of economic efficiency and rationalisation applied to agriculture, especially to farming systems. For some the application has been unusually long-delayed, i.e., farming as a sector has been backward in terms of its organisation, compared say to manufacturing; for others it is seen as disruptive but inevitable — the price of progress. Generally, however, agro-technology is matched with progress in the agricultural sector, particularly as a means to achieving the desired end of production efficiency. The latter is the key economic objective and its realisation, through increased scale of operation and concentration of activity is measured in terms of higher land and labour productivity. Once again, these economic principles are commonly espoused — by western agri-businessmen, agricultural scientists, bureaucrats and 'successful' farmers, *and* by socialist technocrats, agricultural planners, and collective and state farm managers. If the former have to defend their viewpoint against concerns over Jeffersonian ideals, the latter find it easy to justify in support of large-scale integrated state-operated agricultural systems.

In operational terms, the common inputs of agro-industrial technology and the doctrine of economic efficiency combine to transform modern agriculture through four common processes, namely:

1. *Increases in size of the production unit;* widespread efforts to achieve economies of scale at the farm level and involving a marked reduction in number of farms, including already large, collectivised units.

2. *Specialisation in production;* a shift away from mixed farming towards larger, uniform cropping regimes and an especial emphasis on large-scale, single-type livestock operations.

3. *Intensification of capital inputs;* greatly increased investment per farm and farm worker, both in land and buildings (i.e. scale), livestock (specialisation), and secondary inputs through the application of

mechanical and chemical energy, to facilitate increased output and substitution of human labour.

The three processes identified above operate primarily at the farm level with a combination effect on its transformation in industrial terms. In addition, however, a highly significant fourth process operates more generally within the agro-food system, namely,

4. *Integration* of farm production with other parts of the total agriculture and food system. Operationally, integration strengthens assembly-line linkages between existing input supply and the processing and distribution sectors of agribusiness and farm-level production. The chief mechanism is that of vertical integration, applied either by a capitalist corporate or co-operative structure, or by socialist collectivisation, particularly the new Agro-Industrial Complexes. The latter are a logical extension of the scale of the collective model (CPSU, 1961), the former reflects the breakdown of independence of action within the capitalist farm sector which facilitates (many argue, necessitates) take-over by the more highly centralised supply and processing sectors. In both cases, because of the level of sectoral concentration, integration tends to be imposed upon rather than generated by the farm level.

Each of the four processes, but particularaly their combined effect, results in a more streamlined and efficient input-output sequence to the model (Figure 4.2), with the overall positive results; output achieved at lower unit cost of production and maintenance of an adequate and 'cheap food' supply to the urban consumer. However, there are problems or impacts connected with industrialisation. The most obvious is that it operates on the basis of a continually decreasing set of production units (and producers), and in a capitalist context, very much on the basis of 'the survival of the fittest' (Vogeler, 1981). In addition, although unit costs may be lowered, agricultural output may be more costly in its reliance upon increasing amounts of non-renewable inputs, i.e., its energy efficiency is increasingly in question (Berry, 1977; Pimental *et al.*, 1975). In a broader sense it may represent a final shift away from the intimate relationship between man and nature which is part of the agrarian ideal.

Impacts of Agricultural Industrialisation

Agricultural industrialisation produces some very direct results in terms of change in farm operation. Application of the common processes —

increased scale, capital intensity, specialisation and integration — gives rise to measurable characteristics (see Table 4.1). These provide the basis for economic evaluation of the process, with success measured in strictly economic terms, particularly increases in productivity. A major criticism is that 'success' measured in terms of fewer farms and farmers, producing more per farm and utilising greater amounts of energy, ignores the cost in both human and ecological terms.

Industrialisation results in a range of impacts (Figure 4.2(c)). To a large extent these are regarded as externalities to the economic performance of agriculture, but they are tangible and often affect society outside the agricultural sector. In contrast to the broad scale at which industrialisation has been described (and is often justified) many impacts tend to look small and to pose predominantly local management problems. On the other hand, they tend to be widespread in occurence and incremental in nature. In addition, they are symptomatic of the possibility of more fundamental problems attendant on industrialisation and what some critics have diagnosed as a malignant state of affairs.

Environmental Impacts

From the time of transition from a subsistence to a market orientation, there have been alterations in traditional agricultural land-man relationships; the interruption of local cycling of materials, a gradual divorce of man from his intimate ecological relationship and direct energy input. Industrialisation, however, has vastly increased the pace and scale of each such change.

Environmental impacts may be distinguished as to ecological or landscape amenity effects, but both are directly related to changes in scale, intensity and specialisation. Ecological impacts can be seen on both the cultivated and remaining 'natural' ecosystem components of the farm landscape. Remnants of the natural ecosystem (woodlots, hedgerows, etc.) are falling victim to widespread farm and field enlargement to meet the needs of an increased scale of mechanisation (Westmacott and Worthington, 1974). Use of heavy machines, chemically supported cash cropping of grains, oilseeds, etc., contribute to increased damage to soil structure and fertility and to its actual loss through water and wind erosion. In turn, increased run-off, the product of improved drainage, but including excess fertilisers and pesticides, contributes to pollution of streams and groundwater, eutrophication of lakes and damage to many natural ecosystems; the quantities of organic wastes generated by large-scale specialist livestock operations lead to similar pollution problems (Bangay, 1976).

Table 4.1: Selected Measures of Change in Agricultural Systems of Developed Countries

Country	Time Period	Decrease in no. of farms		Increase in ave. farm size		Decrease in agric. employ	
		Loss '000	% Loss	Gain ha.	% Inc.	Loss '000	% Loss
Australia	1960–80	73	29	67	41	n.a.	—
Austria	1960–80	94	23	3	32	880	73
Belgium/Luxem.	1959–80	86	47	7	67	161	58
Canada	1951–80	305	49	105	93	248	49
Denmark	1960–82	84	43	10	64	192	52
Finland	1959–81	163	43	5	67	473	66
France	1955–80	999	47	13	85	3,348	65
Germany (West)	1960–82	854	53	9	105	2,205	61
Ireland	1961–82	132	37	8	58	197	50
Italy	1961–80	2,088	49	4	95	3,642	55
Japan	1960–80	1,395	23	n.a.	—	6,480	54
Netherlands	1950–80	82	39	5	45	287	54
New Zealand	1960–80	6	8	n.a.	—	n.a.	—
Norway	1959–80	79	40	n.a.	—	88	35
Spain	1962–80	1,068	36	n.a.	—	2,439	53
Sweden	1960–82	153	57	18	135	283	55
Switzerland	1955–80	106	51	10	105	137	50
United Kingdom	1960–80	227	48	34	85	162	23
USA	1960–80	1,360	34	67	61	2,307	40
Bulgaria	1960–81	1	70	15,899	261	1,478	62
Czechoslovakia	1960–81	11[a]	84	3,047	542	509	35
GDR	1961–82	13[a]	75	1,087	303	517	37

Hungary	1960–82	1	46	1,840[a]	138	790	42
				2,041[b]	248		
Poland	1960–82	c. 4[a]	35	272[a]	96	1,189	18
		c. 500[c]	15		—		
Romania	1960–80	1	16	n.a.	19	3,020	50
USSR	1959–82	+14[a]	+217	579	23	n.a.	—
		28[b]	52	4,141			

Notes: a. State; b. Collective; c. Private.
Source: OECD: 1973–75; 1983; World Atlas of Agriculture, 1966 Statesman's Yearbook: Europa Yearbook: Selected national studies.

Other fundamental ecological problems that have been raised include the switch from farming based on organic renewable energy to high-level reliance on non-renewable fossil fuels; the increasingly narrow genetic base which may leave agriculture susceptible to pest and disease outbreaks of epidemic proportions and is a denial of the basic strength of an ecosystem that derives from its diversity (Mooney, 1979); and finally concern over the essential nutritional quality of crops produced under chemical cultivation, and of livestock products from animals raised under the stress of 'factory farming' methods (Hill and Ramsay, 1976).

Many traditional, and even recently formed agricultural landscapes contribute to what is regarded as rural landscape amenity. Visual and aesthetic values may include farm and field layout, buildings of traditional style and materials and even the presence of grazing livestock. The effects of industrialisation including farm and field amalgamation, mono-cultivation, housed livestock and the construction of utilitarian farm buildings, have a tendency to contribute to loss of amenity, including a significant transformation of the rural milieu. In addition, there are problems of air (odour) and water pollution and the propensity for mechanical-chemical farming to produce landscapes unsuitable for alternative countryside uses, including both outdoor recreation and nature conservation (Advisory Council for Agriculture, 1978).

Socio-economic impacts

Agriculture is, above all, a human system, and socio-economic impacts are of major significance. They too have increased as a result of industrialisation. Some distinctions may be made between impacts on individuals in the farm population, those affecting the function of rural communities, and broader impacts on the viability of society in general.

Industrialisation generally means a drastic change in the way-of-life of the farm population. The success of industrialisation is commonly measured in terms of the decreased numbers of farms required and actual reductions in farm population (see Table 4.1). To survive in a capitalist setting the individual farmer has to be competitive and adopt a purely business approach. Not only may this literally pit him against other farmers (even his neighbours) but it carries the increased pressure of technical and financial expertise. Many people have been made redundant and forced to abandon farming, contributing to widespread off-farm migration; many others have taken up off-farm employment and operate only on a part-time basis, not least because of the competitive pressures of cost-price relationships (Vogeler, 1981).

Fewer numbers and increasing specialisation tend to isolate the farmer socially, even within rural society, which may accentuate the stresses attendant on individual management responsibility (McGhee, 1984). This may be somewhat less in co-operatively based societies, but there too enormous losses of farm personnel have occurred. Integration into a contracting or wage-labour situation may lead to problems of loss of individual identity. The latter shows indications of being a particular problem in socialist agriculture (and on large corporate farms) where the majority work only within a narrow speciality, and where there is little personal commitment to the enterprise (Enyedi, 1976a).

Farmer, farm family and farm worker losses contribute to changes in rural communities. Rural settlement functions change in response to the overall reorganisation of agricultural supply, production and processing (Lonsdale and Enyedi, 1984). Increased scale and integration have resulted in major reductions in the size and number of rural service settlements in capitalist countries, with attendant concern over structural problems of declining and aged populations and the maintenance of basic social and economic services, including employment, retailing, education, health and transportation (Rural Community Councils, 1978). Collective farm organisation, especially that based on larger village settlements, may provide a better base on which to preserve rural community, but here also there is evidence of the problems of providing alternative employment, and cleavages between farm and non-farm (commuter) segments of the population (Enyedi, 1976a).

Another major impact is that of increased regional differentiation, especially the problem of areas made marginal by changes in agriculture. By its nature, industrial agriculture emphasises physical and other locational advantages, especially the facility whereby land may be farmed extensively and with machines; as Berry has put it: All land has been divided into two parts; that which permits use of large equipment, and that which does not (Berry, 1977, p.33).

The latter tends to fall out of agriculture. Consequently, a widespread by-product of industrialisation is the increased relative disadvantage experienced by areas poorly endowed in terms of the requirements of agricultural industrialisation i.e., of low productive capacity and/or isolated from markets. In each developed country, economically viable agriculture is becoming concentrated on a smaller area while the marginal area is enlarged, exacerbating problems of social and economic hardship and of regional disparity.

Political Responses

The impacts of agricultural industrialisation have resulted in some

political response, albeit usually of a piecemeal nature. Specific en-
viromental concerns have led to investigation, policy formulation, even
some legislation and regulation in such areas as the application of
agricultural chemicals, agricultural waste disposal and even to control
the alteration of the vegetational and built elements in the landscape.
Concern relating to both physical and human impacts has generated both
official and unofficial political support and there have been specific cases
of conflict between those promoting agricultural and those promoting
non-agricultural land uses. However, in many cases the speed of technical
and economic change within the agricultural sector, which is only poorly
understood by those outside, and a general separation of the majority
from a rural-agricultural setting have meant very little by way of a concerted response to change or its impacts.

In socialist countries planning for rural areas is seen as part of an
overall national planning function and has included attempts to increase
rural employment, income and opportunity, and to upgrade rural services (Enyedi, 1976a). However, environmental impacts are usually given
less priority and it is hard to assess the real impact on the populations
involved. Some capitalist countries have attempted to develop and apply
a comprehensive approach to rural development including attempts to
rationalise the impacts on marginal farming through subsidies, land bank-
ing or retirement schemes and to maintain rural community structure
through provision of health, education and transportation facilities, but
no consensus on the needs of industrialising areas *per se,* seems to exist
(Koutaniemi, 1977).

Having noted both the process and impacts of agricultural industrialisation in overall terms, the final part of the chapter seeks to classify modern
agriculture in terms of types of industrialisation and to discuss some
national examples within each type.

A Preliminary Typology of Agricultural Industrialisation

The classification proposed (Figure 4.3) is both preliminary and largely
qualitative. Whereas the processes of agricultural industrialisation have
produced measurable changes, a few of which have been assembled for
a selection of countries in Table 4.1, no thorough quantitative typological
analysis has been undertaken. What is needed is a comprehensive
assembly of current data similar to the 1960 World Census of Agriculture
(FAO 1969) or the *World Atlas of Agriculture* (1969), which together
provide a baseline from which to measure industrialisation. Nevertheless,

Figure 4.3: A Preliminary Typology of Industrialised Agriculture

TYPE	CHARACTERISTICS	EXAMPLES
'pure' Socialist	*State ownership of agricultural supply, protection & processing.*	*U.S.S.R.*
	Collective farm units; large and undergoing consolidation.	*Bulgaria* *Czechoslovakia* *G.D.R.*
	Integration through Agro-Industrial Complexes; some already established.	*Hungary* *Romania*
	Total government control of market demand & price structure.	
'pure' Capitalist	*Private ownership of agricultural supply, production & processing.*	*U.S.A.* *Canada*
	Individual farms predominant; becoming fewer, larger; widespread adoption of corporate model.	*Australia*
	Integration through relationships between farms & oligopolistic agribusiness; includes contracts, vertical integation.	
	Prices according to supply and demand; limited government intervention but support for economies of scale and business model.	
Cooperative	*Private ownership of agricultural supply, production & processing; co-ops active in each sector.*	*Scandinavian countries.* *Austria* *Switzerland*
	Individual farm units; gradual size increases, but economies of scale via cooperative pooling of output.	*New Zealand*
	Integration through cooperative structures; farmer ownership and decision making.	
	Government support for co-op structure; co-op/farmer -- government negotiation on production & prices.	

there is a body of national studies dealing with aspects and problems of industrialising agriculture which supports the tentative classification (OECD, 1975). It should be noted, of course, that it is over-simplified and that in many countries industrialisation does not encompass the whole agricultural system.

While it is emphasised that industrialisation is the dominant process underlying agricultural change in all developed countries, the differential nature of such factors as farm tenure, the ownership of supply and processing facilities, and the political decision framework relating to supply of and demand for agricultural output combine to produce at least three broad types of industrialised agricultural systems (Figure 4.3).

1. The 'pure' Socialist model is the most consistent, especially in terms of scale and overall integration, with the state controlling the whole 'assembly line', including the total supply and demand situation.
2. The 'pure' Capitalist type, in contrast, is largely in private hands, albeit with some government intervention and implicit support for economic rationalisation. Corporate style integration, however, is still only piecemeal and divisions exist between supply, production and processing levels, particularly between predominant individual farm production and oligopolistic and conglomerate control of supply and processing.
3. Co-operative, represents a third, intermediate type. The name reflects the fact that in certain countries the traditional involvement of farmer-based co-operative organisations at several levels of agricultural organisation has persisted. Although these countries are actively pursuing a reduction in numbers of individual farms and farm workers, and programs of capitalisation, specialisation and integration, the co-operative mechanism, rather than the state or corporation is utilised to develop and apply a policy of rationalisation of the production system and its output.

Socialist Industrialised Agriculture

The countries involved are basically the Warsaw Pact or European Comecon nations,[1] and the dominant model is collectivisation, initiated by the USSR and expanded across Eastern Europe within the post-war decade. Russian agriculture at this time was dominated by the so-called cooperative, actually true collective farms (*kholkoz*) with a smaller set of generally larger state farms (*sovkhozes*). Various combinations of this state system were established in Bulgaria (1945), Romania (1945), Czechoslovakia (1951 ff.), East Germany (GDR) (1952 ff.), Albania (1956), and Hungary (1956) where they now dominate, despite continued existence of private small-holdings (e.g. Hungary) which are economically significant in disproportion to their size. In contrast, collectivisation was only partially achieved in Poland and Yugoslavia. In Poland the system remains sharply divided between the socialist sector now dominated by state farms which occupy about 20 per cent of the farm area, but with a larger share of investment and output of grain and livestock, and a private sector still divided into nearly 3 million small farms, which despite efforts to integrate (some 90,000 'Agricultural Circles') remain essentially outside the industrial sector. The Poles identify the structure of the private sector as a major problem with respect to increased productivity but seem to have stopped attempts at complete collectivisation (Rajtar, 1977). Similarly, in Yugoslavia which opted out of the soviet bloc in 1948 a mixed system of similar proportions causes problems of agricultural development.

Within the seven comprehensively socialised systems there are both common and distinctive aspects to the industrial structure. Since collectivisation, there has been a general trend to further consolidation, particularly a major reduction in the number of collectives (less in the case of state farms), and huge increases in size of individual farms. At the same time there have been reductions in the agricultural labour force of up to 70 per cent between 1960 and 1982 which have reduced this component to levels similar to those prevailing in capitalist nations. There are distinctions, however, which suggest two, possibly three sub-classes.

Countries outside the USSR were still in relatively early stages of applying collectivisation in the 1950s, and in several, notably Czechoslovakia and GDR, many thousands of small collectives were the norm. Since then, both countries have undertaken enormous consolidation, resulting in convergence with countries such as Hungary, Bulgaria and Romania where initial units were larger and fewer and where consolidation has been more measured. In each country, except Bulgaria, the reduced farm numbers are still weighted heavily towards collective rather than state enterprises, although the proportion of the latter has increased.

In contrast, in the USSR a trend which began in the 1950s was for a major reduction in collectives but an increase in state farms. Initially, this reflected activity opening up new farming areas, including the 'virgin lands' in soviet Asia and areas in Siberia. But of greater significance has been the increased application of the economies of scale and specialisation to the whole system through the promotion of Agro-Industrial Complexes (AICs) (CPSU, 1961). The latter seek a planned integration of the whole agro-food system at the regional level and involve the consolidation (or establishment) of regional supply and processing facilities to serve larger production units and emphasising specialist production. The AIC model which was developed in newly settled areas has been increasingly applied to older regions such as the Ukraine and Byelorussia, as well as to Bulgarian agriculture (Shotski, 1979; Enyedi, 1976a). One result has been to reduce the number of collective farms to less than 30 per cent of the 1950 total. Meanwhile the number of state farms has increased fourfold and now represents over 80 per cent of the number of collectives. In Bulgaria the system has concentrated to less than 300 farms.

The situation in the USSR seems to reflect a very highly centralised system of agricultural planning. It has been part of a major effort to mechanise agricultural production and to develop an intensive livestock sector. Reports conflict as to effectiveness, contrasting poor harvests, increased grain imports, and problems of supply and servicing of machines, animal health, worker alienation and low productivity, with assertions of its success in releasing labour and improving consumer supplies. Despite problems, the USSR remains the world's largest grain producer and is attempting to develop agriculture consistent with the 'pure socialist' industrial model.

Presently, in some contrast, the agricultural systems in Hungary, and GDR appear less centralised and more flexible in terms of decisions made at an individual farm level. In Hungary, for example, individual farms evidence innovations which range from importation of western systems (corn-hog feeding from the US) (Dohrs, 1982) to local diversification and even marketing arrangements for meat, eggs, fruit and wine with countries ouside the socialist bloc. On the other hand, rigid structures appear dominant in Czechoslovakia and Romania, and the latter remains backward in terms of mechanisation and labour substitution.

Conditions of rural life, i.e., housing, services, were traditionally poor throughout much of eastern Europe. Collectivisation and agricultural industrialisation have had a mixed impact. A predominantly village population was initially retained as the labour force and there were often severe restrictions on migration to the cities. Subsequently the size of many nucleations has provided the basis for improved services (roads, transport, water, sewers, electricity, etc.) and policies of the decentralisation of manufacturing. The latter has become important as agriculture has shed its labour requirements and many farm villages have transformed into either small manufacturing centres or dormitory settlements for non-farm workers. It is difficult to gauge whether the changes have redressed traditional rural-urban disparities, but there is evidence of a general improvement in social and educational facilities, and many rural families have a larger living space than their urban counterparts (Enyedi, 1976b); Rajtar, 1977). No overall assessment exists of the environmental impacts of socialist agriculture, but observation suggests that the treeless cropland areas are liable to soil erosion, and agro-industrial facilities dot the landscape without regard for aesthetic considerations. Recent global assessment of soil erosion suggests that the USSR constitutes a severe case (Brown and Wolf, 1984) and other reports suggest widespread pollution from livestock operations and agricultural processing plants.

Despite structural problems, including marginal areas, especially in upland and northerly zones, and land and labour productivity generally below levels achieved by capitalist agriculture, there is little sign of weakening of the socialist industrial model, rather attempts to make it more efficient through planning based on the application of agro-industrial technology and economic rationalisation. Although private plots are tolerated in terms of their valuable contributions of vegetables, fruit, etc., they are not being enlarged and solutions to the problems of the major anomalous Socialist state, Poland, tend to be for radical reorganisation of the small-scale, fragmented private sector into co-operative/collective units.

Capitalist Industrialised Agriculture

The primary distinctions between socialist and capitalist agriculture remain the matter of ownership of the means of production and the decision mechanisms with respect to the nature and disposal of production. To that extent, industrialisation in all capitalist countries remains distinguishable from the socialist type. Here, however, the distinction between so-called 'pure capitalist' and 'co-operative' types, is made in the belief that key distinctions within the capitalist decision framework result in somewhat different forms of, and responses to, industrialisation. Tentatively, the 'pure capitalist' type is found most clearly in the USA and the 'co-operative' type in Scandinavia. Another 15 or so countries, the majority of members of the OECD and identified by the World Bank as 'Industrial Market Economies' (World Bank, 1982) lie at varying points on a spectrum between. Generally, Canada, Australia, the United Kingdom, West Germany, and France tend to 'pure capitalist' although government and farmer involvement and the impact of European Community agricultural policy interferes somewhat in the three latter countries. Countries with stronger 'co-operative' structures, and where farming remains smaller in scale, include Austria, Ireland, the Netherlands, New Zealand and Switzerland.

The 'Pure Capitalist' Type

Despite the pervasive involvement of government in agriculture, from basic scientific research, through provision of credit, to policies which affect both production and income, US agriculture, both the farm and non-farm sectors, is dominated by private enterprise. There are still over two million individual farms, but the reduction from the peak of 6.5

million continues and, in reality, output and income are heavily concentrated in a much smaller minority of very large and/or capital intensive operations, concentrated in regions of highest physical capability and/or market accessibility (Schnepf, 1979). Although corporate farming in its fullest sense has not developed in the manner predicted in the 1950s (Higbee, 1958), seemingly because of problems of maintaining labour productivity and by low returns which diverted capital to more attractive investments, nevertheless corporate control is strongly exercised by high levels of concentration (oligopolies) within the supply and processing sectors (Crittenden, 1981; Greenhouse, 1984).

On the other hand, as Gregor has noted (1982), widespread farm industrialisation has occurred through the application of the range of secondary inputs by individual farm operators, resulting in very high levels of capital intensity. Inputs have included land purchase or rental and average farm size has increased substantially, but use of machines, fertilisers and chemicals, etc. has been even more pronounced. The process has been very competitive, but those who were successful achieved huge output and high income associated with the world's most productive system in the 1970s (*Time*, 1978). Recently, however, there have been signs of economic stress; high interest rates, a stagnant domestic market, and the impact on exports of an inflated US dollar have combined to place the farmer in a severe cost-price squeeze. Falling incomes and high indebtedness have resulted in a large number of bankruptcies, and some questioning of the industrialisation model. Farm input suppliers have also been hard hit, especially the machinery industry, which has been reduced to less than half peak output and employment, and even processors have been affected by the high costs of production and weak markets (Greenhouse, 1984). The latter undoubtedly have the resources to weather the situation but the severe impacts on farmers will continue if, as experts predict, depressed conditions persist until the 1990s (Robbins, 1985).

There are strong similarities between US agriculture and that in Canada and Australia, the other major nineteenth centrury pioneer areas which followed the US model (Scott, 1982; Troughton, 1982); these include drastic reductions in farm numbers, heavy emphasis on the top 40 per cent of producers located in key regions, capital intensification and the current problems associated with the cost-price squeeze. However, there are certain factors which act to modify the situation. Although corporate interests in both countries enjoy oligopolistic control of farm inputs and the processing and distribution chain, there is very little corporate farming.[2] Farmer movements in all countries were organised to try to offset the influence of big business, but whereas in the US large private firms dominate commodity marketing (Morgan, 1979), in Canada and Australia, governments have assumed control of key export items, i.e., wheat in Canada and wheat and wool in Australia. Consequently, the Canadian Wheat Board, for example, rather than one of the five multinational private companies, services its 106,000 farm members and helps to maintain their cooperative organisations (the prairie wheat pools). On the other hand, the number of prairie farms has decreased drastically and the average size is now close to 300 ha. (Wilson, 1981). However, in Canada both federal and provincial governments have instituted product marketing boards which now affect a majority of output. The 'supply management' approach seeks not only to regulate output, but to allocate production to farmers on a quota system which tends to slow down the competitive reduction in numbers of producers. Contrasts exist between Canada and the US in levels of concentration in such areas as egg and broiler production, where unregulated economies of scale in the US have concentrated production in a handful of major producers.

In both the US and Canada agriculture is affected by a policy of cheap food to the consumer (i.e. cheap in terms of proportion of disposable income spent but not actual cost of production). This tends to maintain the cost-price squeeze on the farmer from both government and the processing sector. The latter is organised into strong lobby groups which are often more effective than the farm organisations, whose members remain uncertain as to which economic philosophy and market arrangement is in their best interest (Warnock, 1978). The agricultural policy of the European Community, in contrast, serves to articulate the Western European 'farm lobby' and results in prices for production above those in a 'pure market' situation.

The Co-operative Type

Co-operative in this context is quite distinct from socialist agriculture where the term is used synonymously for collective. Rather, co-operative farming is based solidly on retention of individual ownership by the farm operator, but utilises co-operative action to achieve economies of scale in purchasing and particularly in the marketing and processing of output. It developed first in, and remains solidly part of, the agricultural systems of the Scandinavian countries (Denmark, Finland, Norway and Sweden) and plays a strong role today in many Western European countries with the exception of the United Kingdom, as well as in New Zealand and Japan. Co-operative agriculture pre-dates industrialisation and in many ways is antithetical to its tendencies. In Scandinavia, for example, as late as the 1960s, efforts were under way in Denmark and Finland to increase the numbers and to safeguard the existence of new small farms (Kampp, 1975; Varjo, 1977). Nevertheless, in all developed areas, Scandinavia included, the pressures of technological change and economic rationalisation have been felt, which since 1960 have resulted in major reductions in farm numbers and employment, and major increases in capital investment. However, co-operation has adapted, and in two key areas alters the nature and impact of industrialisation. At the farm level, although numbers have been reduced, the reductions have been less severe than the limited size of the farm system might have suggested; compare Scandinavia's remaining 560,000 farms with only 250,000 in the UK. In addition, in marketing and price regulation, farmer co-operatives retain a large measure of control versus private agribusiness. These effects are realised by the power of co-operative farm and marketing organisations which are recognised by governments as the legitimate agents of the farm system and with whom joint negotiations are carried out (Knudsen, 1977; OECD, 1975, 1983).

The impacts of industrialisation in capitalist countries are very varied and the responses not easily categorised within the economic framework. The environmental and ecological impacts of increased scale of operation, field size, intensive livestock and applications of fertilisers and chemicals have been noted in most areas, concerns voiced, and some attempts make at mitigation. Some impacts are necessarily exaggerated by the scale of operation, hence problems of soil erosion on huge arable areas are most evident in North America where potential disaster situations have been noted (Bertrand, 1980; Sparrow, 1984), and in new extended cropland areas in northern France and southern England, where there has been public outcry over destruction of the previous, more vegetated agricultural landscape (Westmacott and Worthington, 1974). The major problem of pollution from intensive livestock has affected many areas, especially landscapes where small farms and natural watercourses are easily overwhelmed by large quantities of effluent.

In the case of erosion and waste management, the problems, though severe, may be capable of solution without threatening the industrial model. Soil erosion control methods such as rotation, strip cropping, undersowing, etc. can be reintroduced, while a major innovation in the form of minimum or zero tillage has already been utilised on half the row crop acreage in the USA (Phillips *et al.*, 1980). This latter approach is interesting in that it reduces the machine energy inputs into cropping, but may increase the requirements for pesticides, another non-renewable resource item.

The value placed on the farm landscape as part of an aesthetic or recreational amenity resource varies widely, being generally of greater concern in Western Europe than in North America (Munton, 1983). Consequently, in the latter areas very little is being done to counter vegetation removal, uniformity in building and allowance of agro-industrial structures on farms and in open countryside; these are matters subject to more stringent controls especially in the United Kingdom, Germany and Scandinavia.

The social impacts of industrialisation on capitalist rural communities have been very significant, especially outside metropolitan regions and in areas where a dispersed farm pattern has become even more so. Reduction of the farm population has had drastic impacts in rural service communities in economically viable agricultural areas (e.g. US mid-west, Canadian prairies) as well as in marginal areas of agricultural withdrawal. Despite recognition of rural problems neither the US nor Canada is involved in comprehensive rural development planning. Both countries tried to mitigate the severe problems in marginal areas in the 1960s but these programmes were never applied to industrialising regions.

Similarly in the European context, although more consistent attention has been paid to marginal rural areas, both by individual governments and through the Regional Policy of the European Community, the impacts on industrialising regions *per se* have been of less concern. Many countries have applied subsidies to help maintain agriculture in poorer areas (e.g., Finland, Italy, Norway, UK) and to retain rural populations and communities (Koutaniemi, 1977). Studies in the UK have documented the decline of rural communities in core farming areas (i.e., lack of services and employment) (Rural Community Councils, 1978) but cause and effect have not been linked in any planning package. Similarly, until very recently, the environmental impacts have been dealt with in two separate contexts; agricultural operations and landscape conservation. However, the realisation that the impacts of industrialisation have occured very rapidly over the last 20–25 years and the recognition of some overall effects have been subject to a lag effect. Given growing concerns on the part of large segments of the population in areas such as wildlife and historic preservation and for landscape amenity, especially in Western Europe, mitigating policies may evolve relatively quickly (Munton, 1983).

On the other hand, industrialisation has been attacked in more fundamental ecological and economic terms and has been identified as a potential problem beyond its present extent in developed countries.

Industrialisation has proceeded furthest in North America which has produced some of the most intense criticism. For some, industrialised agriculture is the latest and most inhumane, exploitive stage in the use of the North American environment and is contributing to the rapid demise of its only stable rural base, the yeoman or family farm. The agent of industrialisation is the corporate institution, whose essentially exploitive nature is based on the short-term goal of efficiency measured in terms of profit. Two quotations summarise these concerns:

> The cost of this corporate totalitarianism in energy, land and social disruption will be enormous. It will lead to the exhaustion of farmland and farm culture. Husbandry will become an extractive industry; because maintenance will entirely give way to production, the fertility of the soil will become a limited unrenewable resource. (Berry, 1977, p.47)

> An agriculture cannot survive long at the expense of the natural systems that support it and that provide it with models. A culture cannot survive long at the expense of either its agricultural or its natural sources. (Ibid.)

Wider concerns have also been expressed, namely the diffusion of industrialised agriculture beyond the developed world. There have long been elements of modern agriculture within traditional economies, the colonial plantation with its capital, know-how and international trading linkages being the most obvious. Many former plantation crops remain important sources of revenue for developing countries and the problems of a dual economy with its investment emphasis on export crops, to the detriment of food production, is well known. There are suggestions that forms of industrialisation are amplifying such tendencies in key areas, particularly with the development of a large-scale export sector competing for resources with small-scale peasant farmers. The situation seems to be most acute in Latin America where industrial approaches have been applied by large landholders (*latifundista*) who devote large areas to production of beef, soybeans, and even flowers, for export, while denying land to millions of smallholders (*minifundista*) (Place, 1982).

Conclusion

The thesis advanced here is that modern agricultural systems show strong evidence of change through a set of processes and based on beliefs that are creating various forms of industrialised agriculture. The transformation is relatively recent in origin but very pervasive. Despite its impacts, some of which are harmful and most of which act to break down existing agrarian structures, the transformation seems likely to be permanent because it emanates not from within agriculture, the minority sector, but from a dominant urban-industrial sector acting to reduce the differences between agriculture and the rest of the industrial economy.

Notes

1. Albania withdrew from Comecon in 1962 and from the Warsaw Pact in 1969.
2. Although some Australian farms are among the largest in the world their output is based on very low productivity lands and they are not corporate farms in an industrial sense.

References

Advisory Council for Agriculture and Horticulture in England and Wales (1978) *Agriculture and the Countryside*, UK
Bangay, G.E. (1976) *Livestock and Poultry Wastes in the Great Lakes Basin*, Social Science Series No.15, Environment Canada, Burlington, Alberta
Berry, W.E. (1977) *The Unsettling of America: Culture and Agriculture*, Sierra Club Books, Avon, New York
Bertrand, A.R. (1980) 'Overdrawing the nation's research accounts', *Journal of Soil and Water Conservation*, 35 (3), 109–15
Brown, L.E. and Wolf, E.C., (1984). 'Soil erosion: quiet crisis in the world economy', *Worldwatch Paper 60*, Worldwatch Inst., Washington
Cobbett, W. (1830) *Rural Rides*, Penguin English Library Edition (1967)
Committee for the World Atlas of Agriculture (1969) (G. Medici, ed.) *World Atlas of Agriculture*, Istituto Geografico de Agostini-Novara, Italy (4 vols.)
CPSU (Community Party of Soviet Union) (1961) *Proceedings of XXIInd Congress*, Moscow, p. 383
Crittenden, A. (1981) 'More and more conglomerate links in US food chain', *New York Times*, 1 Feb, p. E3
Dohrs, F.E. (1982) 'The modernization of agriculture as a factor in rural transformation: Hungarian and American analogues' in G. Enyedi and I. Volgyes (1982) pp. 199–212
Enyedi, G. (ed.) (1976a) *Rural Transformation in Hungary*, Hungarian Academy of Sciences, Budapest
Enyedi, G. (ed.) (1976b) *Agrarian-Industrial Complexes in the Modern Agriculture*, Proceedings, IGU Symposium, Budapest
Enyedi, G. and Volgyes, I. (eds.) (1982) *The Effect of Modern Agriculture on Rural Development*, Pergamon, New York

FAO (UN) (1969) *World Census of Agriculture 1960*, Rome

Fussell, G.H. (1965) *Farming Technique from Prehistoric to Modern Times*, Pergamon, Oxford

Goldschmidt, W. (1947) *As You Sow: Three Studies in the Social Consequences of Agribusiness*, Free Press, Glencoe, Illinois

Greenhouse, S. (1984) 'Farm equipment hits a trough', *New York Times*, 11 Nov.

Grigg, D.B. (1974) *The Agricultural Systems of the World*, Cambridge University Press, Cambridge, (Part 1)

Gregor, H.F. (1974) *An Agricultural Typology of California*, Geography of World Agriculture 4, Akademiai Kiado, Budapest

Gregor, H.F. (1981) 'Large-scale farming as a dilemma in US rural development, IGU Commission on Rural Development, Fresno, California (mimeo)

Gregor, H.F. (1982) *Industrialization of U.S. Agriculture: An Interpretive Atlas*, Westview, Boulder, Colorado

Harlan, J.R. (1975) *Crops and Man*, American Society of Agronomy, Madison, Wisconsin

Higbee, E. (1958) *American Agriculture: Geography, Resources, Conservation*, Wiley, New York

Higbee, E. (1963) *Farms and Farmer in an Urban Age*, Twentieth Century Fund, New York

Hightower, J. (1975) *Eat your heart out*, Vantage, New York

Hill, S.B. and Ramsay, J.A. (1976) Limitation of the energy approach in defining priorities in agriculture, St. Louis (mimeo.)

Jensen, E. (1937) *Danish Agriculture: Its Economic Development*, Schulz Forlag, Copenhagen

Johnson, G.L. (ed.) (1972) *The Overproduction Trap in US Agriculture*, Johns Hopkins, Baltimore

Jones, E.L. (ed.) (1967) *Agriculture and Economic Growth in England, 1650–1815*, Methuen, London

Kampp, Aa.H. (1975) *An Agricultural Geography of Denmark*, Geography of World Agriculture 5, Akademiai Kiado, Budapest

Knudsen, P.H. (ed.) (1977) *Agriculture in Denmark*, Agricultural Council of Denmark, Copenhagen

Koutaniemi, L. (ed.) (1977) *Rural Development in Highlands and High-Latitude Zones*, Proceedings IGU Symposium, Oulu, Finland

Kramer, M. (1980) *Three Farms*, Atlantic, Little Brown, Boston/Toronto

Lonsdale, R.E. and Enyedi, G. (eds.) (1984) *Rural Public Services: International Comparisons*, Westview, Boulder, Colorado

Lydolph, P.E. (1979) *Geography of the USSR*, Wiley, Elkhart Lake, Wisconsin

McCarty, H.H. and Lindberg, J.B. (1966) *A Preface to Economic Geography*, Prentice Hall, Englewood Cliffs, New Jersey

McGee, M. (1984) *Women in Agriculture*, OMAF, Toronto

Mitchell, D. (1975) *The Politics of Food*, Lorimer, Toronto

Mooney, P.R. (1979) *Seeds of the Earth: A Private or Public Resource?* US Institute of Food and Development, London/Ottawa

Morgan, D. (1979) *Merchants of Grain*, Penguin, New York

Munton, R. (1983) 'Agriculture and conservation: what room for compromise' in A. Warren and F.B. Goldsmith (eds.) *Conservation in Perspective*, Wiley, London, pp. 353–73

OECD (1973–75) *Review of Agricultural Policies* (24 countries), Paris

OECD (1975) *Review of Agricultural Policies: General Survey*, Paris

OECD (1983) *Review of Agricultural Policy in OECD Member Countries 1980–82*, Paris

Phillips, R.E., Blevins, R.L., Thomas, G.W., Frye, W.W. and Phillips, S.H., (1980) 'No-tillage agriculture', *Science, 208*, 1108–13

Pimental, D., Hurd, L.E., Bellotti, A.C., Forster, M.J., Oka, I.N., Sholes, O.D., and Whitman, R.J., (1975) 'Food production and the energy crisis' in P.H. Abelson (ed.), *Food, Politics, Economics, Nutrition and Research*, AAAS, Washington, pp. 121–7

Place, S. (1982) *The Underdevelopment of Costa Rica*, IGU Rural Development Commission, Aracaja, Brazil (mimeo.)

Rajtar, J. (ed.) (1977) *Village and Agriculture*, Polish Academy of Sciences, Warsaw

Robbins, W. (1985) 'Hardship in farms forecast in study', *New York Times*, 29 Jan.

Rural Community Councils (1978) *The Decline of Rural Services*, Standing Conference, London

Schnepf, M. (ed.) (1979) *Farmland, Food and the Future*, SCSA, Iowa

Scott, P. (1982) *Australian Agriculture*, Geography of World Agriculture 9, Akademiai Kiado, Budapest

Shotski, V.P. (1979) *Agro-Industrial Complexes and Types of Agriculture in Eastern Siberia*, Geography of World Agriculture 8, Akademiai Kiado, Budapest

Smith, E.G. Jr (1980) 'America's richest farms and ranchers', *Annals of the Association of American Geographers*, 70 (4), 528–41

Sparrow, H.O. (1984) *Soil at Risk: Canada's Eroding Future*, Senate of Canada, Ottawa

Symons, L. (1972) *Russian Agriculture: A Geographic Study*, Halsted Press, New York

Time 'The new american farmer', 6. Nov, pp. 52–64

Tranter, R.B. (ed.) (1978) *The Future of Upland Britain*, CAS Paper 2, Reading, UK (2 vols.)

Troughton, M.J. (1982) *Canadian Agriculture*, Geography of World Agriculture 10, Akademiai Kiado, Budapest

United Kingdom (1979) *Annual Review of Agriculture 1979*, HMSO

Varjo, U. (1977) *Finnish Farming*, Geography of World Agriculture 6, Akademiai Kiado, Budapest

Vogeler, I. (1981) *The Myth of the Family Farm*, Westview, Boulder, Colorado

Warnock, J. (1978) *Profit Hungry*, New Star Books, Vancouver

Westmacott, R. and Worthington, T. (1974) *New Agricultural Landscapes*, Countryside Commission, Cheltenham

Wilson, B. (1981) *Beyond the Harvest: Canadian Grain at the Crossroads*, Prairie Books, Saskatoon

Wittfogel, K.A. (1956) 'The hydraulic civilizations' in W.L. Thomas (ed.), *Man's Role in Changing the Face of the Earth*, University of Chicago Press, Chicago, pp. 152–64

World Bank (IBRD) (1982) *World Development Report 1982*, Washington, DC

5 GOVERNMENT AGRICULTURAL POLICIES

I.R. Bowler

Government intervention is greater in agriculture than in any other sector of the economy and is an international phenomenon. Consequently 'the state' has become one of the most important factors shaping both the structure of farming and the location of agricultural production. However, the degree of intervention varies: from the state control of inputs to and outputs from agriculture in centrally planned, communist countries such as the Soviet Union; through an intermediate mixture of private and state control of the factors of production as in Yugoslavia and Poland; to the provision of income supports for farms largely in private ownership as, for example, in the EEC, and the relatively low level of supply control practiced in capitalist-oriented countries such as Canada. Moreover, the degree and style of intervention are not fixed but wax and wane in all countries according to changing economic, political and ideological conditions.

So vast is the literature on agricultural policy that some limitation must be placed on this discussion. Consequently, attention has been focused on agricultural policies in countries with developed, market economies and democratic systems of government. Agricultural policies in communist countries have been examined by writers such as Clayton (1984) and Mollett (1984). Even so, governments can intervene in agriculture either directly or indirectly, the latter including fiscal (taxation), international trade and monetary policies, as well as regulations on such features as property inheritance, land development and the monopoly powers of large, agribusinesses. These indirect policies are excluded from the discussion although their importance should not be discounted. On the direct policies, reference has been made only to those books and journals readily accessible to the reader and written in English.

Neither a country by country, nor a policy by policy approach has been adopted in the discussion. Reviews of this type are provided periodically by OECD (1983), while FAO has recorded agricultural legislation throughout the world annually since 1965 (FAO, 1965). Rather, a framework is advanced to meet a division in the literature between the domains of 'political co-ordination mechanisms' and 'market co-ordination mechanisms' (Hagedorn, 1983). The former concerns the way in which policy is formulated and draws on the work of political scientists and sociologists, while the latter looks at policy instruments through the research of economists and geographers. To date, few attempts have been made to bring together these two domains within the literature on agricultural policy.

Deriving the Goals of Agricultural Policy

Agricultural policy is rooted in the value system of each society. Values can be described as 'pattern principles' for, in the words of Parks (1968), they guide human action in the sense that people have beliefs, preferences and aspirations regarding what is desirable. Most countries have their underlying value systems formally expressed in a constitution, although they become translated into more tangible beliefs on how the economy or society should be organised. Kazereczki (1971) shows how values for the economic system can refer to the improvement of living standards, economic progress, balanced growth or economic security. Equity values, especially for agriculture, commonly involve concepts such as income parity with the non-farm sector and income stability. A common set of values can be identified for most democratic countries; indeed, a supra-national organisation such as the EEC depends on the existence of a convergence of societal values. Even so, values can be given a different order of priority in each country, especially when interpreted for particular sectional interests such as agriculture. In countries with a large farm population, agriculture can be viewed as a distinct and unique sector of society. The term 'rural fundamentalism', as discussed by Tracy (1964, p.369), is given to the most extreme expression of rural values. This philosophy includes the views that the ideals and virtues of an agrarian society are superior to the values held by an urban society, and that a vigorous democracy is supported by a stratum of owner-operated family farms (Anderson, 1972). Of course, the influence of rural values has declined in step with the farm population while, as reviewed by England *et al.* (1974), there has been a levelling of values between the city dweller and the countryman. Nevertheless, many policy goals were set decades ago when rural values were more significant; they continue into the present under the force of inertia. Moreover, there exists a residual, broad-based sympathy for the farm population even in urban-oriented societies.

Unfortunately the free market does not provide an optimum single solution for meeting the varied values and desires expressed within a society. Thus the political process enables different values to be translated into more specific policy goals, with governments intervening in the market to regulate economic forces towards desired ends. Sectional interests tend to seek conflicting goals and, moreover, enjoy varying political power in the pursuit of those goals. In agricultural matters, farm groups, consumers, environmentalists (Lowe and Goyder, 1983) and taxpayers commonly conflict in setting goals for agriculture; the government often has to apply the concept of a 'national interest' in giving more weight to one group compared with another. The interpretation of 'national interest' varies with the prevailing macro-economic environment as well as the ideology of the ruling political party, a theme developed for the United Kingdom by Bowler (1979, pp. 46–54). As Englebert (1968) stresses, with policy-making a balance of interests, the resulting compromises do not always make a policy adequate to the values or beliefs being pursued.

Most observers agree that agriculture has been able to exert an influence on this policy-making process out of proportion to its contribution to the economy or employment structure of most countries. 'Interest group' theory has been developed to explain how governments respond to the demands of effective interest groups, although Wilson (1977) has argued that there is no single common reason for the political power of agriculture and that different factors explain the influence of farmers in different countries. Even so, two bases to agriculture's political power can be discerned. First, farmers are organised into a number of cohesive and vociferous interest groups. Some interests are narrowly focused around a single product like sugar-beet or milk; others are broadly based and maintain a unity of purpose amongst a range of farmers. The latter type of group is often recognised by governments as representing the whole agricultural sector in consultations over agricultural policy. Their power can be increased by maintaining a non-partisan stance in party politics. The considerable influence of the National Farmers' Union in the United Kingdom, for example, can be traced to these factors (Self and Storing, 1962) but, as Mackintosh (1970) points up, this is not always in the interest of society at large. Policy goals and alternatives tend to be discussed in secret and are not sufficiently open to investigation and examination by the general public. The close relationship between farm groups and civil servants also leads to accusations that ministries of agriculture have been 'captured' by farm interests. Ministers of agriculture are seen as representing farm interests at the expense of consumers and taxpayers on government decision-making bodies, such as the British cabinet. In some countries, however, the power of the farm lobby has been diluted either by the fragmentation of groups, as in the

United States (Bonnen, 1968), or by too close an identification with a political party, as in Italy. In the latter case, the Cottivatori Diretti is a Christian Democrat organisation while Confagricultura, representing large landowners, is attached to the Liberal Party.

The second basis of agriculture's influence lies in its direct political voting power. There is conflicting evidence, however, concerning the homogeneity of the agricultural vote. Research by Talbot and Hadwiger (1968), for example, suggests that farmers in the United States switch parties more often than any other occupational group thus generating political leverage with the two main political parties. On the other hand, Knoke and Long (1975) were unable to find evidence of mobility in the agricultural vote in the presidential elections of the 1950s and 1960s. A more direct influence has been identified in Sweden by Cohen (1975). Farmers traditionally support the Centre Party; following the elections of 1973, when the Centre Party increased its holding of Parliamentary seats, agricultural policy was redirected from reducing farm output to encouraging the production of exportable products such as wheat, pork and oilseeds. The agricultural vote has enjoyed a similar pivotal role in the balance of power between political parties in West Germany (Andrlik, 1981). In Britain, however, farm families are relatively few in number and form part of a broader 'rural vote' (Benyon and Harrison, 1962). Nevertheless, it is often asserted that land-owning Members of Parliament, and individuals in the House of Lords, exert a considerable influence on agricultural matters. Recent research in the United States by Welch and Peters (1983), however, was unable to confirm that private interests influence an individual's voting behaviour on agricultural policy.

Clearly, farmers are not the only groups in society with a vested interest in agricultural policy. Guither (1980), for the USA, has examined the broader range of 'food lobbyists' who influence policy either overtly or covertly, and this perspective reflects a general shift in emphasis in the literature away from the farm and to the whole food chain. Indeed Halcrow (1977) has demonstrated why it is now more accurate to speak of 'food policy' rather than 'agricultural policy'. Agricultural policy has been 'externalised' to embrace wider issues such as national standards of nutrition (Centre for Agricultural Strategy, 1979), alternative forms of agricultural production, such as organic farming (Rushefsky, 1980) and global food supplies (Balaam and Carey, 1982). With the increasing significance of other interest groups in policy formation, such as agribusiness, food aid agencies and consumers, agriculture has lost its 'unique' place in agricultural policy. In addition, recent research has emphasised the growing influence on policy-making of bureaucrats

(Page, 1985), specialist advisers and civil servants within national treasuries (Infanger *et al.*, 1983). Recent redirections in agricultural policy within both the EEC and the United States, for example, can be traced to constraints placed on the budgets available to fund policy measures (Goodenough, 1984).

All of these considerations are made more complex for a supra-national organisation such as the EEC. Here national interests are advanced and defended in the Council of Ministers which forms the decision-making body. Only the European Commision defends a 'Community interest' (Williamson, 1983) and national interests dominate. Farm groups have had to adapt to take account of this policy-making structure by forming a federation of farmers' unions — COPA (Comité des Organisations Professionnelles Agricoles). COPA represents the farm lobby on the complex system of influential committees that advise and form policy within the EEC. In addition COPA co-ordinates national farm groups in their dealings with individual national ministers of agriculture (Averyt, 1977). These ministers then represent agriculture in, as Buksti (1983) puts it, the 'bread-and-butter' decision-making of the Council of Ministers.

Usually, the long-run goals of agricultural policy are framed within general legislation; the relevant legislation of the EEC, for example, is the 1957 Treaty of Rome (Articles 38–47). Most countries have developed similar policy goals reflecting the convergence in their value systems, but, following Self and Storing (1962, p. 218), goals can be subdivided into two groups. In one group are goals that seek 'equity' for those employed in agriculture, usually defined in terms of farm income. In a second and larger group are goals that require an increasing utility from agriculture in its contribution to the economy. The various goals can be redefined in increasingly specific terms so that a hierarchy of goals is achieved. These are not described here but can be seen in Figure 5.1. Although the goals are similar, countries vary in the priority given to each goal. The EEC, together with Austria, Norway, Sweden and Switzerland, for example, gives a higher priority to the goals of food security (self-sufficiency) and consumer price stability than to the costs of farm support programmes.

For those who wish to study agricultural policy, these goals pose a number of immediate problems. For example, several are conflicting. Thus, protection of the environment is the most recently developed policy goal but higher ecological standards in farming tend to increase production costs and retard the rate of technological advance (Timmons, 1972). Another problem is caused by the very general terms in which goals are couched. 'Equitable remuneration', for instance, is rarely quantified so

Figure 5.1: The Goals of Agricultural Policy

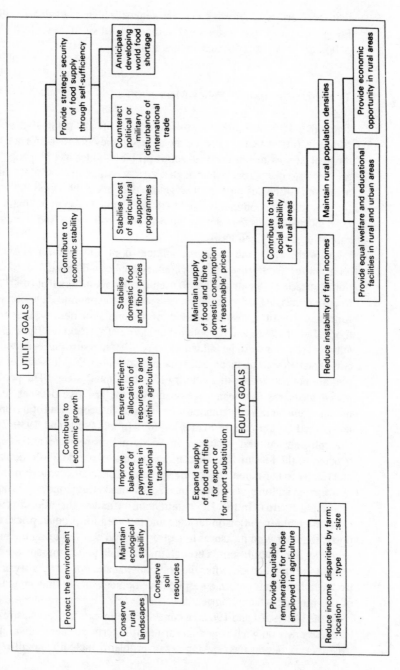

that there are few objective criteria against which to evaluate policy. For geographers, the problems are compounded since it is rare for any policy objective to be stated in spatial terms.

Types of Agricultural Policy Measure

Through the political process, governments create and manipulate policy measures (instruments) to reach the agreed policy goals. Unfortunately the analyst now meets a second set of problems since any single policy goal tends to be approached through a variety of measures, while a single measure is often designed to reach more than one goal. Price supports, for example, are commonly used both to orientate supply and maintain farm incomes. Thus the relationship between a policy goal and a measure cannot be determined easily.

As with values and goals, a common 'pool' of measures exists for all developed countries, but each country applies a distinctive combination of measures. Heidhues (1970) usefully lists six factors that contribute to policy measure differentiation: the stage of economic development, farm size structure, resource endowment, population density, degree of food self-sufficiency and the political influence of agriculture. The societal value system should be added to this list; measures acceptable to counteract low farm incomes in socialist-oriented Sweden, for example, are not acceptable in a country like the United States. The pool of policy measures has been developed over the last five decades, each measure reflecting a contemporary explanation by economists of the problems faced by agriculture (Tracy, 1964). In the 1920s and 1930s, for example, agricultural problems in Western Europe were initially attributed to the loss of foreign markets, then to slack domestic demand and finally to inefficient marketing systems. Consequently tariff protection for agriculture was supplemented by marketing controls through marketing boards. In the 1950s, attention turned to the impact of the technological revolution on farming and the need for product price supports, together with measures to restrict supply, so as to bring the farm economy into equilibrium. Through inertia, such policy measures often remain in force decades after the conditions under which they were introduced have changed, and despite developments in our understanding of the problems facing agriculture.

In the late 1950s and 1960s, a considerable research effort was made by economists on what is now known as 'the farm problem'. In reality it is not one but an interlocking set of problems including small-farm structural inefficiency, the long-run oversupply of markets, a cost-price squeeze on producers, annual variability in production and farm incomes and, centrally, the persistence of low incomes in agriculture. Hill and Ingersent (1977, p. 106) have been able to identify five separate theories to account for the last element only — low farm incomes. A different policy measure is relevant to each explanation be it the declining income elasticity of demand for food with economic growth, the continued application of output-increasing farm technology despite falling real product prices, the low level of management in agriculture, or the increased rates of product yield to human resources from investment in education. The contemporary consensus view of the farm problem has been reviewed by Josling (1974) and Bowler (1979, pp. 13–20) and is not repeated here. The central thrust of the explanation, however, interprets the farm problem as a function of the dynamic process of economic development leading, in the words of James (1971), to a maladjustment in the allocation of resources to agriculture. In this interpretation, labour, as the most mobile factor of production, needs to be transferred to other sectors of the economy until equilibrium is restored between the demand for and supply of agricultural products. As labour is withdrawn, a farm structure of fewer but larger holdings is created — the 'structural solution' to the farm problem. Consequently, a range of measures has been developed in most countries to promote the mobility of labour out of agriculture including retirement pensions, financial aid for farm amalgamation and retraining schemes for non-farm work. Farm incomes have been protected by price supports while the process of transition takes place.

Unfortunately the scale of mobility by the farm population has been insufficient to resolve the farm problem, not least because the level of income support has been so generous as to remove the economic incentive for leaving agriculture (Aall, 1979). Nevertheless, 'psychic' values (Bellerby, 1956) have been identified as a reason for labour immobility together with loss of status, independence and kinship ties (Gasson, 1969). Moreover, the availability of alternative employment outside agriculture has contracted in the 1980s with structural unemployment in most developed economies, while the application of output-increasing farm technology has over-ridden the beneficial effects of a reduction in the farm population. Since the need to reduce the farm population varies regionally, agricultural policy has also been 'externalised' by becoming dependent on regional rural development programmes. As Clout (1984) has eloquently argued, only these programmes seem capable of generating needed non-farm employment in rural regions.

Table 5.1: A Typology of Agricultural Policy Measures

Measures designed to influence:

(i) *The growth and development of agriculture*

: subsidies on the cost of inputs (fertilisers, machinery, buildings), drainage, ploughing
: preferential taxes
: preferential credit interest rates
: financial assistance to co-operatives or syndicates for the joint purchase of inputs
: financial assistance to agricultural educational extension and research establishments

(ii) *The terms and levels of compensation of production*

: demand management
— advertising
— produce grading schemes
— multiple pricing
: supply management
— land retirement
— import quotas
— tariff protection and import quotas
— marketing quotas
— production quotas
— buffer stock schemes
— target/threshold prices with intervention (support) buying
— export subsidies
— bilateral/multilateral commodity agreements
: direct payments
— deficiency payments
— fixed rate subsidies/production grants
— direct income support payments

(iii) *The economic structure of agriculture*

: financial assistance for land consolidation schemes
: farm amalgamation
: land reform schemes
: retirement and discontinuation pensions/grants
: retraining schemes
: financial assistance to co-operatives for production or marketing
: inheritance laws (and taxation)
: farm wage determination

(iii) *The environmental quality of rural areas*

: pollution control measures
: land zoning ordinances
: flood control measures
: landscape conservation controls
: wildlife conservation grants

Table 5.1 classifies the profusion of policy measures that have been developed to meet the farm problem. There are four headings: the growth and development of agriculture, the economic structure of agriculture, the environmental quality of rural areas and the terms and levels of compensation of production. The first three headings cover measures that approach utility goals for agriculture, while the last heading relates to the equity goals of agricultural policy. The mechanisms of the various measures are not discussed here; rather, the reader should turn to one of the numerous texts where such details are given (Fennell, 1979; OECD, 1983).

Evaluating the Impact of Agricultural Policy Measures

Most of the research on 'market co-ordination mechanisms' (Hagendorn, 1983) accepts the goals as given and considers various means of attaining them. Outside geography, researchers and policy advisers have been concerned with either the aggregate performance of the agricultural sector or groups of farms by size or type of production. Spatial studies have encountered particular problems. For example, in addition to, and probably because of, the absence of spatial objectives, few policy measures have an explicitly spatial or regional dimension. This feature limits the scope of geographical analysis. Further, policy-related statistics are rarely available for individual farms, while spatially disaggregated data at best are usually provided only for large administrative areas, so masking more detailed spatial information. Spatial considerations rarely influence the data collected by government agencies and this limits the type of geographical question that can be asked and answered in relation to agricultural policy. Taken together, these features explain why comparatively few evaluations of agricultural policy have been published by geographers compared with agricultural economists.

Measuring the Costs of Agricultural Protection

One of the principal functions of agricultural policy is to protect farming from the full force of the free market. The direct costs of such protection are highly visible and are usually published annually by government agencies. The direct cost of the CAP in 1983/84, for example, was approximately £10 billion. Deficiency payments, direct income supplements and grant aids for investment all fall into this category, and such costs are borne by taxpayers.

A second set of costs are indirect and less visible. They emanate from the separation of domestic and world markets with higher prices in the former than the latter market. Tariffs and import levies are commonly used to create this separation, as in the EEC. Systems based on import levies impose a welfare loss on low-income and large households, for the consumer bears the higher cost of food. As we will see, larger farms producing the highest agricultural outputs gain most under protective agricultural policy measures so that a resource transfer takes place from the poorest, usually urban, families to the most wealthy rural (farm) families (Castellani, 1977; Howarth, 1971). Deficiency payments, by contrast, exact a relatively low welfare cost since, as Josling and Hamway (1976) show, the cost burden is shared amongst taxpayers approximately in accordance with income.

None of these costs is easily quantified, a central problem being posed by the need to estimate the degree of protection given to agriculture. Strak (1982), for example, has developed analytical measures such as 'nominal', 'adjusted' and 'effective' tariff rates which emphasise the importance of substitution in agriculture, particularly between traded and non-traded farm inputs. A less complex approach has been to compare the valuation of domestic production at domestic prices with the valuation of that production at current import prices (McCrone, 1962). Debate has centred on the selection of appropriate import prices since protectionist policies themselves have an impact on world, or import, prices. At present only residual volumes of agricultural production enter world trade. Most estimates reveal a varying level of protection amongst developed countries both in aggregate and by individual product. Moreover, the degree of protection varies through time so that care is needed both in the choice of commodity for analysis and the time period (Corbet and Niall, 1976). Thus, while the EEC appears to support a highly protected agriculture in aggregate, individual products are protected more heavily in other countries. Japan and Canada, for example, provide more protection to their dairy sectors than does the EEC.

By maintaining domestic prices above those ruling on world markets, productive resources, especially capital, can be retained unnecessarily in agriculture. Quantifying the 'opportunity cost' of this misallocation of resources is also problematic for it requires assumptions on how the capital would be deployed elsewhere in the economy. Over-capitalisation of agriculture has been claimed by Body (1983) for the United Kingdom where it takes the form of over-mechanisation, excessive investment in buildings and infrastructure, and inflated land values. Indeed there is widespread recognition that the value of government price support programmes is often capitalised into the price of farm land rather than raising farm incomes in real terms. In addition, the capital gains are weighted in favour of those owning large farms (Anderson, 1973).

Where supra-national agricultural policies are implemented, as in the EEC, resource transfers take place between countries as well as between sectors of the economy. O'Connor *et al.,* (1983) are one of many research teams who have shown that net food importing countries suffer a negative 'food trade' transfer, whereas net food exporters enjoy a positive transfer. This arises because countries trade their food within the Community (common market) at prices in excess of the world price. Most estimates show the Netherlands, Denmark and Ireland gaining at the expense of Italy, West Germany and the United Kingdom.

Another facet of agricultural protection lies in the field of international trade. Governments increasingly determine the conditions and extent of international trade through bilateral and multi-lateral trade agreements such as GATT, UNCTAD and, in the case of the EEC, the Lomé Conventions. In addition, the protectionist policies pursued by all developed countries significantly reduce the markets available to agricultural exporters, especially developing countries, while export subsidies can undermine other traditional markets. The degree to which trade diversion takes place is open to debate (Knox, 1972) but, as Feltner (1983) shows, current trade patterns depart substantially from those that would be determined purely by comparative advantage. An alternative view is to consider agriculture's import-saving role, especially when a country suffers a balance of payments problem. It is a traditional theme for economists in the United Kingdom (Ritson, 1970) but there is now a wider interest in countries such as Portugal (Lopes, 1975) and New Zealand (Hume, 1961). Once again there are methodological problems in measuring features such as the elasticity of demand for exports and reciprocity coefficients, but the balance of informed opinion in the United Kingdom has favoured an expansion of agriculture (Houston, 1975).

Monitoring the Income Effects of Government Intervention

The primary justification for agricultural policy lies in the protection of farm incomes. Most commentators now agree that income support given through trade restriction to this end is inefficient relative to the same degree of support through a production subsidy or deficiency payment, and that these measures in turn are more costly than direct income payments unrelated to production. Of course when objectives other than raising farm incomes are considered, the economic efficiency of different support measures changes. Josling *et al.*, (1972) show that a tariff system is superior to a subsidy if the objective is to provide a minimum level of income, or a maximum producer price. On the other hand, if the objective is to save foreign exchange, a mixture of deficiency payments, supply limitation and production subsidy is to be preferred.

Most measures aimed at increasing or maintaining farm incomes focus on product prices. Consequently, producers with the greatest volume of agricultural output gain the greatest absolute benefit. Brech (1961), for example, estimated that 50 per cent of the total support given to farmers in the United Kingdom in the 1950s was taken up by the 13 per cent of farms over 61 hectares in size. Indeed a wide range of analyses across many countries and over a variety of price support measures shows the benefits of government intervention to be concentrated on the largest farms (Johnson and Short, 1983). Attempts at limiting the maximum amount of direct support to any farm have achieved only partial success. The legal dispersion of farmland ownership amongst several individuals can readily circumvent such constraints. In addition, the farm lobby often stresses the need to reward the efficiency of larger farms and to support new investment by favourable levels of price support (Albers, 1969). In the EEC, no such limitations exist, and price levels set so as to give a sufficient income to the occupiers of small farms are so generous as to encourage unwarranted production and financial benefits to those on large holdings. The alternative of providing direct income supplements has been resisted in most countries. Koester and Tangermann (1977) outline the arguments of farm groups that income supplements are tantamount to social security benefits, require a bureaucracy for their payment, and remove the incentive to improve farm efficiency. Experience with 'compensatory allowances' under the Less Favoured Areas Directive of the EEC, however, does not support these unduly pessimistic assertions, while all the theoretical analyses demonstrate the superiority of direct payments in supporting farm incomes (Heads, 1974).

Just how effective agricultural policy has been in providing income parity between the farm and non-farm sectors of the economy remains an active area for debate. There are formidable conceptual and statistical difficulties to be overcome, not least in identifying the non-farm group with which to compare incomes. Arguably, any farmer provides skilled manual labour and managerial expertise as well as risk capital in his enterprise. Manual wage rates, therefore, are not necessarily a valid indicator of comparable non-farm incomes, while professional salaries fail to account for the element of capital provision. In addition, calculating farm incomes is fraught with problems (Hill, 1983). Non-wage, family labour should be accounted for, in addition to income-in-kind and the rising capital value of the land. None of these measures is commonly recorded by farm income data networks which, in any event, tend to record more profitable, commercially-oriented farm businesses. More recently, with the significant development of part-time farming, the debate has been extended to consider farm-family income, rather than just farm income, as the relevant indicator for comparison with the non-farm sector. This is an important consideration when farm income levels are used to determine the level of protection to be accorded to the agricultural sector.

Another problem which is masked by looking only at aggregate farm incomes is the extent of income variability within agriculture. Most attention has been devoted to variability by farm size with the conclusion that the 'income gap' between large and small holdings has not been closed by government intervention. In addition incomes vary by farm type (OECD, 1964), physical resource base and even distance from urban markets (Taurianinen and Young, 1976). Consequently there are marked spatial variations in farm income both between countries and between regions in each country (Gillmor, 1971). When support is linked to the prices of particular commodities, it is not surprising that agricultural policy measures can exaceberate rather than ameliorate spatial variations in farm income. Davidson and Wibberley (1956), for example, in a pioneering study, showed that disproportionately more subsidy was paid to lowland farmers in the United Kingdom than to upland farmers, while Henry (1981), for the EEC, has demonstrated the favoured position of farmers in 'northern' regions, in terms of income support, compared with Mediterranean regions of the Community. In the United States, with most money being used to support the prices of corn and wheat, and to a lesser extent cotton and tobacco, funds have flowed towards Iowa, Nebraska and North Dakota (Brunn and Hoffman, 1969).

One broad guide to the income status of farming can be gained by comparing the proportion of the workforce employed in agriculture with the proportion of GNP generated by agriculture. As these values converge, so there is evidence of income parity. Table 5.2 shows that income parity is being approached in only a few countries (Belgium, Canada, United Kingdom) and that the position is worsening in others (United States, Spain, European Community). Indeed farm incomes, in real terms, have fallen in all countries over the past decade, especially in the United Kingdom.

Assessing the Efficacy of Policy Measures

Farm income and national food production goals are approached by manipulating the inputs to and outputs from agriculture, as well as the structure of farming itself. Often these manipulations cause unexpected and undesirable changes in agriculture which require further corrective measures themselves.

Farm Inputs. Capital is the main input considered here, with relevant measures listed in Table 5.1 under the heading 'the growth and

Table 5.2: Indicators of Farm Income

Country	Income Parity[a] GNP/AE		Income Trend 1982/1976 x 100 in real terms
	1970	1980	
Norway	0.47	0.56	85
Belgium	0.84	0.93	75
Canada	0.49	0.71	75
France	0.48	0.49	73
Japan	0.35	0.38	73
USA	0.61	0.35	65
Germany	0.40	0.40	63
Spain	0.41	0.39	63
United Kingdom	0.81	0.74	55
European Community	0.63	0.53	—

Notes: a: A value of 1.00 indicates parity between farm and non-farm incomes.
GNP: gross value added by agriculture as a percentage of gross national product.
AE: agricultural employment as a percentage of total civilian employment.
Source: Calculated from data in OECD, 1983, pp. 148, 152, 162–3.

development of agriculture'. Land and labour are also subject to government intervention but they are discussed under 'farm structure'. Governments influence capital inputs indirectly by regulating fiscal policy on matters such as capital gains tax, allowances set against income tax and estate duty, and also by lending capital at preferential rates of interest. In addition, most governments fund agricultural research and extension workers and provide the finance for education through agricultural colleges. More directly, subsidies can be provided for variable inputs such as fertilisers, machinery and ploughing, while grant aid is commonly available to promote fixed investment in items such as farm buildings, infrastructure (farm roads, electricity, water supply) and field drainage. Grants and subsidies probably encourage unwarranted capital expenditure by farmers at the margin of profitability, and also stimulate over-capitalisation on the largest farms. Folley (1972), for example, found a low rate of return to capital investment in the United Kingdom under an Horticultural Improvement Scheme. Moreover, the input of government-financed capital varies by farm type. Green (1976) shows how drainage, lime and fertiliser grants have been used most extensively in the arable farming areas of eastern England, while Delamarre (1976), for France, demonstrates a high rate of adoption of grant aid for livestock buildings in the départements of Brittany, Nord, Midi-Pyrénées and Pays-de-la-Loire. Closer inspection, however, reveals three dimensions to grant aid (Bowler, 1979, pp. 87–103). Marked spatial inequalities of direct government assistance result from a combination of the initial pattern of adoption (innovativeness), subsequent variations in resistances to the adoption of grant aid by farm size and type, and the amount of grant per farm. Because all three dimensions are least developed in those areas most in need of capital investment, government assistance perpetuates and magnifies existing inter-regional differences in the economic performance of agriculture.

Farm Outputs. Governments attempt to guide farm production by setting production targets and manipulating the terms and levels of compensation of production (product prices). In democratic, market-oriented countries governments tend to guide rather than fix prices so that the 'market' still plays a role, albeit constrained, in the final farm-gate price. Broadly, two methods have been developed for estimating the supply response of agriculture to variations in price (Cowling and Gardner, 1963). On the one hand, regression and autoregressive models have been used at the aggregate market level to analyse time series, often adopting a Nerlove-type partial adjustment, or lagged, model. Such models yield estimates of the supply elasticity in relation to price. On the other hand, linear programming has been applied at the farm-firm level, but with less certain conclusions when results are aggregated for industry-level estimates (Buckwell and Hazell, 1972). Both types of approach demonstrate low supply responsiveness to price changes in the short run. Other factors tend to assume an equal if not over-riding influence on variations in output including the net profitability of the product, the weather and, as demonstrated by Fielding (1965) for wheat production in New Zealand, the profitability of competing farm enterprises (price ratios). These conclusions apply equally to cereals (Hill, 1971), potatoes (Revell, 1974) and milk (Gardner and Walker, 1972).

In the longer term, government manipulation of price can act as a 'signal' to producers while also creating a stable economic environment within which investment decisions for production can take place. However, since prices are also manipulated for their effect on farm income, and large variations in price are considered to be undesirable, the 'signalling' role of any price change is greatly diluted. Moreover, when farmer attitudes to government intervention are examined (Hoiberg and Bultena, 1981), price 'signals' are credited with little impact on decision making. Locational changes in agriculture within the EEC, for example, are a function of the complex interaction of competitive trade within a common market, differential national rates of inflation, and

Figure 5.2: Production Changes under the CAP: (a) Cereals, (b) Fruit and Vegetables (changing share of total agricultural production by value, 1964–1976)

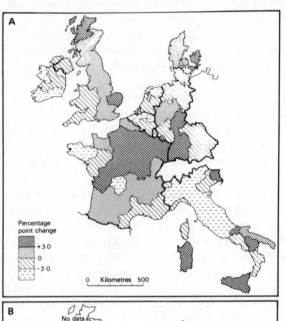

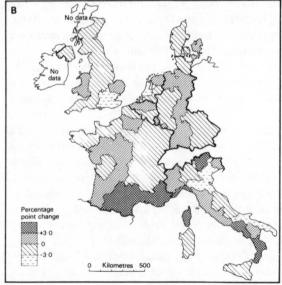

national differences in the 'common' prices fixed for the Community (Bowler, 1985). Figure 5.2, shows the greater regional localisation of cereals production in northern and central areas of the EEC under these processes, whereas fruit and vegetables are increasingly located in the Mediterranean regions.

Of course the management of supply can be achieved by more direct measures which, when used in combination rather than individually, can be particularly effective. Several studies of government programmes to regulate the area under wheat (Morzuch *et al.*, 1980), maize (Houck and Ryan, 1972) and sorghum (Ryan and Abel, 1973) demonstrate the very close control possible when loan rates, support and diversion (set-aside) payments, and area quotas are used together. Evans (1980), for example, shows how the judicious use of prices·increased the area under set-aside programmes through the 1977 Food and Agriculture Act in the United States. Used in isolation, less favourable results are observed. For example, early area allotment (quota) schemes merely shifted crop production (cotton, tobacco) from non-quota to quota land, with an intensification of production on the latter (Hart, 1959). However, new uses had to be found for the land taken out of production; soya beans became a distinctive crop in the Mississippi Valley, for instance, following the allotment of cotton production. Also land retirement (soil bank) schemes have tended to be more effective on marginal, low-yielding land (Hewes, 1967) with undesirable social and economic side-effects owing to the reduced demand for rural services. In addition, when allotment and land retirement schemes remain in force for several years, patterns of production in agriculture become fossilized (Raitz, 1971).

Farm Structure. Governments can intervene in the structure of farming (farm size distribution) in a variety of ways, several of which are discussed in Bowler (1983). In developed countries, governments can intervene in the original pattern of land settlement (Taylor, 1945), through programmes of land reform as examined world-wide by King (1978), and through farm consolidation programmes which have been reviewed by King and Burton (1983). The latter vary from the simplest exchange of parcels of land, as in Belgium, to ambitious programmes of regional rationalisation of the farm structure, as in the Netherlands.

Farm enlargement through amalgamation is arguably the most widespread and significant type of structural change in developed countries in recent years. The rationale of a 'structural solution' to 'the farm problem' was outlined earlier, with governments promoting farm size change through grants and subsidies to cover the legal and operational costs of purchasing additional land. Countries providing 'enabling' financial assistance, such as those forming the EEC, have met with relatively limited success. Consequently, many countries have introduced a land agency charged with buying and selling land to ensure a rational pattern of farm enlargement and to facilitate the process of farm-size change. The impact of such agencies has varied according to their powers of compulsory purchase. Where agencies have to compete on the open market with other buyers, their impact has been limited. France's regional SAFERs (Sociétés d'Aménagement Foncier et d'Etablissement Rural), for example, have been effective only where the rural exodus has been greatest and the land cheapest (Perry, 1969). Similarly, a very haphazard impact has been reported for Farm Enlargement Agreements under Canada's 1965 Agricultural and Rural Development Act (Bunce, 1973). Relatively high levels of structural change, by contrast, have been experienced in Sweden under the influence of the County Agricultural Boards. Whitby (1968) has emphasised the greater powers that the Board can exercise over land transactions compared with other countries.

Most governments also provide voluntary retirement and retraining schemes to speed-up the withdrawal of farm occupiers from agriculture. The results have been disappointing in most countries largely because the financial inducements have not been sufficiently generous. In addition, agricultural support prices have been set at such a high level as to reduce the economic incentive to leave farming (Dams, 1977). Moreover, researchers such as Clout (1975) and Naylor (1976) have revealed a varied spatial impact of retirement programmes, with patterns following differences in age structure, the demand for farmland in the region, the effectiveness of agricultural advisers and even the influence of local lawyers. In general, retirement pensions appear to have been taken mainly by those who would have retired in any event.

Finally, mention must be made of two other forms of government intervention in the farm structure. One is the control exercised over the rate and location of the conversion of farmland to non-agricultural uses. Land development controls have been reviewed recently by Dawson (1984) and have been the subject of intense debate both in the United Kingdom (Rogers, 1978) and North America (Furuseth and Pierce, 1982; Hoffmann, 1982). The agricultural impact of land zoning, on the other hand, has received surprisingly little attention, although Fielding (1962) has shown how zoning has protected dairy farming in California. The second type of intervention is even more direct and concerns the state, or public, ownership of agricultural land. Again this theme has received little attention although McKinley (1955) and Clark (1981) have shown the potentialities for such research.

Conclusion

This chapter offers a broader perspective on agricultural policy than is commonly adopted in most discussions. It stresses four structural elements in the relationship between government and agriculture — societal values, policy goals, policy measures and policy impacts — as well as the processes that link them together. These processes include the translation of values into goals, the matching of measures and goals, and the transmission of measures into agricultural impacts. Taking a broad view of the last process, it is difficult to disagree with Johnson's (1973, p.23) five conclusions on agricultural policy: the policies of developed countries have had adverse effects on world trade, especially for developing countries; policies have induced often unwarranted increases in farm production; domestic product prices have been distorted in relation to import prices; significant welfare and budgetary costs often result from protectionist policies; and the income objective is being reached only for certain sections of the farm population. This discussion emphasises the need to examine 'non-market' (i.e. political) failure (Pasour, 1980) in the formation of agricultural policy, rather than the failure of policy measures which have tended to be the main focus of attention. In particular, policy goals in the past have been defined mainly for a narrow farm interest. But, with the decline of the political power of the agricultural sector, and the 'externalisation' of agricultural policy to include issues such as the food chain, international food trade and aid (Tarrant, 1982), and regional rural development, a new order of priority in policy goals is emerging. Particularly influential are the costs of agricultural intervention and the environmental consequences of policy measures (Bowers and Cheshire, 1983).

In terms of the geography of agriculture, the reader should now be aware of the varying emphasis given to different policy goals and measures amongst developed countries, not least because of the varying agricultural context of each country. A distinction has to be drawn, for example, between agricultural policy in land-scarce countries such as Japan and the United Kingdom, and those countries, such as Australia and the United States, where land is in more ample supply. Policies also vary between countries where small-scale, relatively inefficient producers predominate, as in Italy. In addition, individual policy measures can have one of three influences on agriculture. First, measures applied unselectively to agriculture have a spatially varying effect according to farm size and type. The occupiers of large farms, for example, are in a stronger economic position than their competitors to respond to financial inducements as regards input subsidies or investment grants, and can adjust the balance of their enterprises more readily in accordance with the price 'signals' offered by governments. Similarly, general price changes evoke spatially varying production responses according to regional variations in the combination of farm enterprises. Secondly, measures that are applied selectively to particular regions can evoke selective responses both in production and patterns of investment. Selective measures, however, are adopted with reluctance by governments for there are political, as well as practical, problems of providing financial aid to some farmers while denying it to others. Thirdly, some measures have an inertia rather than change effect on farming. High levels of price support and protection, for example, tend to diminish the rate of farm-size change, while production quotas or allotment programmes can fossilise patterns of production and obstruct the evolution of farming systems. Ultimately, though, agricultural change in the context of policy-making is merely a means to several ends; for this reason a broad perspective is needed to comprehend fully the nature of government intervention in agriculture.

References

Aall, P.R. (1979) 'Agricultural modernisation in England, France and Denmark' in D.J. Puchala (ed.), *Food, Politics and Agricultural Development*, Westview Press, Boulder, Colorado, pp. 45–72

Albers, W. (1969) 'How agricultural policy can contribute to ensuring a livelihood for farmers' in *Proceedings of the Annual Conference*, Gesellschaft für Wirtschafts und Sozialwissenschaften des Landbaues, Giessen, Munich

Andrlik, E. (1981) 'The farmers and the State: agricultural interests in West German politics', *West European Politics, 1*, 104–19

Anderson, J.R. (1973) *A Geography of Agriculture in the United States South East*, Akademiai Kiado, Budapest

Anderson, W.J. (1972) 'Policy alternatives for Canadian agriculture', *Canadian Farm Economics, 7*, 1–7

Averyt, W.F. (1977) *Agropolitics in the European Community: Interest Groups and the CAP*, Praeger, New York

Balaam, D.N. and Carey, M.J. (1982) *Food Politics: the Regional Conflict*, Croom Helm, London

Bellerby, J.R. (1956) *Agriculture and Industry: Relative Incomes*, Macmillan, London

Benyon, V.H. and Harrison, J.E. (1962) *The Political Significance of the British Agricultural Vote*, Exeter University, Exeter

Body, R. (1983) *Agriculture: the Triumph and the Shame*, Temple Smith, London

Bonnen, J.T. (1968) 'National policy for agriculture and for rural life' in R.J. Hildreth (ed.), *Readings in Agricultural Policy*, University of Nebraska Press, Lincoln, Nebraska, pp. 95–107

Bowers, J.K. and Cheshire, P. (1983) *Agriculture, the Countryside and Land Use,* Methuen, London

Bowler, I.R. (1979) *Government and Agriculture: a Spatial Perspective,* Longman, London

Bowler, I.R. (1983) 'Structural change in agriculture' in M. Pacione (ed.), *Progress in Rural Geography,* Croom Helm, London

Bowler, I.R. (1985) *Agriculture under the Common Agricultural Policy, A Geography,* Manchester University Press, Manchester

Brech, R. (1961) 'Economics of agriculture — an economic analysis of the efficiency of British agriculture over time', *Journal of Agricultural Economics, 14,* 446–71

Brunn, S.D. and Hoffman, W.L. (1969) 'The geography of federal grants-in-aid to States', *Economic Geography, 45,* 226–38

Buckwell, A.E. and Hazell, P.B.R. (1972) 'Implications of aggregation bias for the construction of static and dynamic linear programming supply models', *Journal of Agricultural Economics, 23,* 119–34

Buksti, J. (1983) 'Bread and butter agreement and high politics disagreement', *Scandinavian Political Studies, 6(NS),* 261–80

Bunce, M. (1973) 'Farm consolidation and enlargement in Ontario and its relevance to rural development', *Area, 3,* 13–16

Castellani, L. (1977) 'The Community's agricultural policy', *Rivista di Politica Agraria, 24,* 13–20

Centre for Agricultural Strategy (1979) 'National food policy in the UK, *CAS Report 5,* University of Reading, Reading

Clark, G. (1981) 'Public ownership of land in Scotland', *Scottish Geographical Magazine, 97,* 140–6

Clayton, E. (1984) *Agriculture, Poverty and Freedom in Developing Countries,* Macmillan, London

Clout, H.D. (1975) 'Structural change in French farming: the case of the Puy-de-Dôme', *Tijdschrift voor Economische en Sociale Geografie, 66,* 234–45

Clout, H. (1984) *A Rural Policy for the EEC?,* Methuen, London

Cohen, M.H. (1975) 'New directions in Swedish agricultural policy', *Foreign Agricultural Economic Report* 104, Economic Research Service, USDA, Washington

Corbet, H. and Niall, J. (1976) 'Strategy for the liberalisation of agricultural trade', in B. Davey *et al.,* (eds.) *Agriculture and the State,* Macmillan, London, pp. 221–36

Cowling, K. and Gardner, T.W. (1963) 'Analytical models for estimating supply relations in the agricultural sector: a survey and critique', *Journal of Agricultural Economics, 15,* 439–50

Dams, T. (1977) 'Agricultural structural policy in the framework of regional development', *Canadian Journal of Agricultural Economics, 25,* 90–8

Davidson, B.R. and Wibberley, G.P. (1956) *The Agricultural Significance of the Hills,* Wye College, Ashford, Kent

Dawson, A.H. (1984) *The Land Problem in the Developed Economy,* Croom Helm, London

Delamarre, A. (1976) 'Les bâtiments modernes d'élevage en France', *Revue Geographique des Pyrénées et du Sud-Ouest, 47,* 139–58

Dunman, J. (1975) *Agriculture: Capitalist and Socialist,* Lawrence and Wishart, London

England, J.L. *et al.,* (1979) 'The impact of a rural environment on values', *Rural Sociology, 44,* 119–36

Evans, S. (1980) 'Acreage response to the target price and set-aside provisions of the Food and Agriculture Act of 1977', *Agricultural Economics Research, 32,* 1–11

FAO (1965) *Food and Agricultural Legislation* (biannually), United Nations, Rome

Feltner, R.L. (1983) 'Structuring agriculture within a free market oriented economy in the western world', *Agrekon, 22,* 6–11

Fennell, R. (1979) *The Common Agricultural Policy of the European Community,* Granada, London

Fielding, G.J. (1962) 'Dairying in cities designed to keep people out', *Professional*

Geographer, 14, 12–17

Fielding G.J. (1965) 'The role of Government in New Zealand wheat growing', *Annals of the Association of American Geographers, 55,* 87–97

Folley, R.R.W. (1972) 'Investment in practice: with special reference to grant-aided production of crops under glass', *Journal of Agricultural Economics, 23,* 25–34

Furuseth, O.J. and Pierce, J.T. (1982) 'A comparative analysis of farmland preservation programmes in North America', *Canadian Geographer, 26,* 191–206

Gardner, T.W. and Walker, R, (1972) 'Interactions of quantity, price and policy: milk and dairy products', *Journal of Agricultural Economics, 23,* 109–18

Gasson, R.M. (1969) 'Occupational immobility of small farmers', *Occasional Paper* 13, Department of Land Economics, University of Cambridge

Gillmor, D.A. (1971) 'The regional pattern of farmers' income in the Republic of Ireland', *Irish Geography, 6,* 307–15

Goodenough, R. (1984) 'The great American crop surplus: 1983 solution', *Geography, 69,* 351–3

Green, F.H.W. (1976) 'Recent changes in land use and treatment', *Geographical Journal, 142,* 12–25

Guither, H.D. (1980) *The Food Lobbyists: Behind the Scenes of Food and Agri-politics,* Lexington Books, Lexington, Massachusetts

Hagedorn, K. (1983), 'Reflections on the methodology of agricultural policy research', *European Review of Agricultural Economics, 10,* 303–23

Halcrow, H.G. (1977) *Food Policy for America,* McGraw Hill, New York

Hart, J.F. (1959) 'Cotton goes west in the American South', *Geography, 44,* 43–5

Heads, J. (1974) 'Raising farm incomes: transfer payments versus output restrictions', *Canadian Journal of Agricultural Economics, 22,* 31–9

Heidhues, T. (1970) *Agricultural Policy: the Conflict between Needs and Reality,* University of Gottingen, Gottingen

Henry, P. (1981) 'Study of the regional impact of the Common Agricultural Policy', *Regional Policy Series* 21, Commission of the European Communities, Brussels

Hewes, L. (1967) 'The Conservation Reserve of the American Soil Bank as an indicator of regions of maladjustment in agriculture', *Weiner Geographische Schriften* 24/29, 331–46

Hill, B. (1983) 'Farm incomes: myths and perspectives', *Lloyds Bank Review, 149,* 35–48

Hill, B.E. (1971) 'Supply responses in crop and livestock production', *Journal of Agricultural Economics, 22,* 287–93

Hill, B.E. and Ingersent, K.A. (1977) *An Economic Analysis of Agriculture,* Heinemann, London

Hoffman, D.W. (1982) 'Saving farmland, a Canadian programme', *Geojournal, 6,* 539–46

Hoiberg, E.O. and Bultena, G.L. (1981) 'Farm operator attitudes toward govenment involvement in agriculture', *Rural Sociology, 46,* 381–90

Houck, J.P. and Ryan, M.E. (1972) 'Supply analysis for corn in the US, *American Journal of Agricultural Economics, 54,* 184–91

Houston, A.M. (1975) 'Agricultural expanison, import substitution and the balance of payments: a comparative study', *Journal of Agricultural Economics, 26,* 351–65

Howarth, R.W. (1971) 'Agricultural support in Western Europe', *Research Monograph* 25, Institute of Economic Affairs, London

Hume, L.J. (1961) 'Import saving and the balance of payments', *Australian Journal of Agricultural Economics, 5,* 23–33

Infanger, C.L. *et al.,* (1983) 'Agricultural policy in austerity: the making of the 1981 Farm Bill', *American Journal of Agricultural Economics, 65,* 1–9

James, G. (1971) *Agricultural Policy in Wealthy Countries,* Angus and Robertson, Sydney

Johnson, D.G. (1973) *World Agriculture in Disarray,* Macmillan, London

Johnson, J.D. and Short, S.D. (1983) 'Commodity programmes: who has received the benefits?', *American Journal of Agricultural Economics, 65,* 912–21

Josling, T.E. (1974) 'Agricultural policies in developed countries: a review', *Journal of Agricultural Economics, 25,* 229–264

Josling, T.E. and Hamway, D. (1976) 'Income transfer effects of the Common Agricultural Policy', in B. Davey *et al.,* (eds.), *Agriculture and the State,* Macmillan, London, pp. 180–208

Josling, T.E. *et al.,* (1972) 'Burdens and benefits of farm support policies', *Agricultural Trade Paper* 1, Trade Policy Research Centre, London

Kazereczki, K. (1971) 'Goals and problems in national planning of agriculture' in E.O. Heady (ed.), *Economic Models and Quantitative Methods for Decisions and Planning in Agriculture,* Iowa State University Press, USA, pp. 377–93

King, R.L. (1978) *Land Reform: a World Survey,* Bell, London

King, R.L. and Burton, S. (1983) 'Structural change in agriculture: the geography of land consolidation', *Progress in Human Geography, 7,* 471–501

Knoke, D. and Long, D.E. (1975) 'The economic sensitivity of the American farm vote', *Rural Sociology, 40,* 7–17

Knox, F. (1972) *The Common Market and World Trade: trade patterns in temperate-zone foodstuffs,* Praeger, New York

Koester, U. and Tangermann, S. (1977) 'Supplementing farm policy by direct income payments', *European Review of Agricultural Economics, 4,* 7–31

Lopes, J. (1975) 'Portugal: farm trade deficit spurs government programmes', *Foreign Agriculture, 13,* 12–13

Lowe, P. and Goyder, J. (1983) *Environmental Groups in Politics,* George, Allen and Unwin, London

Mackintosh, J.P. (1970) 'The problems of agricultural politics', *Journal of Agricultural Economics, 21,* 74–93

McCrone, R.G.L. (1962) *The Economics of Subsidising Agriculture,* George, Allen and Unwin, London

McKinley, C. (1955) 'The impact of American federalism upon the management of land resources' in A.W. McMahon, (ed.) *Federalism, Mature and Emergent,* Doubleday, New York, pp. 305–27

Mollett, J.A. (1984) *Planning for Agricultural Development,* Croom Helm, London

Morzuch, B.J. *et al.,* (1980) 'Wheat acreage supply response under changing farm programs', *American Journal of Agricultural Economics, 62,* 29–37

Naylor, E.L. (1976) 'Les réformes sociales et structurales de l'agriculture dans le Finistère', *Etudes Rurales, 62,* 89–111

O'Connor, R. *et al.,* (1983) 'A review of the Common Agricultural Policy and the implications of modified systems for Ireland', *Broadsheet* 21, Economic and Social Research Institute, Dublin

OECD (1964) *Low Incomes in Agriculture: problems and policies,* OECD, Paris

OECD (1983) *Review of Agricultural Policies in OECD Member Countries 1980–1982,* OECD, Paris

Page, E.C. (1985) *Political Authority and Bureaucratic power: a comparative analysis,* Univeristy of Tennessee Press, USA

Pasour, E.C. (1980) 'A critique of federal agricultural programs', *Southern Journal of Agricultural Economics, 12,* 29–37

Perry, P.J. (1969) 'The structural revolution in French agriculture', *Revue de Géographie de Montreal, 23,* 137–51

Raitz, K.B. (1971) 'The government institutionalization of tobacco acreage in Wisconsin', *Professional Geographer, 23,* 123–6

Revell, B.J. (1974) 'A regional approach to the potato acreage planting decision', *Journal of Agricultural Economics, 25,* 53–63

Ritson, C. (1970) 'The use of home resources to save imports: a new look', *Journal of Agricultural Economics, 21,* 121–31

Rogers, A. (ed.) (1978) *Urban growth, farmland losses and planning,* Rural Geography

Study Group, Wye College, Ashford

Rushefsky, M.E. (1980) 'Policy implications of alternative agriculture', *Policy Studies Journal, 8*, 772–84

Ryan, M.E. and Abel, M.E. (1973) 'Supply response of the US sorghum acreage to government programs', *Agricultural Economics Research, 25*, 45–55

Self, P. and Storing, P. (1962) *The State and the Farmer*, George, Allen and Unwin, London

Strak, J. (1982) *Measurement of Agricultural Protection*, Trade Policy Research Centre, London

Talbot, R.B. and Hadwiger, D.F. (1968) *The Policy Process in American Agriculture*, Chandler, San Francisco

Tam, O. (1985) *China's Agricultural Modernisation*, Croom Helm, London

Tarrant, J.R. (1982) 'EEC food aid', *Applied Geography, 2*, 127–41

Taurianinen, J. and Young, F.W. (1976) 'The impact of urban-industrial development on agricultural incomes and productivity in Finland, *Land Economics, 51*, 192–206

Taylor, G. (1945) 'Towns and townships in southern Ontario', *Economic Geography, 21*, 88–103

Timmons, J.F., (1972) 'Public land use policy: needs, objectives and guidelines', *Journal of Soil and Water Conservation, 27*, 195–201

Tracy, M. (1964) *Agriculture in Western Europe: crisis and adaptation since 1880*, Jonathan Cape, London

Welch, S. and Peters, J.G. (1983) 'Private interests and public interests: an analysis of the impact of personal finance on Congressional voting on agricultural issues', *Journal of Politics, 45*, 378–96

Whitby, M.C. (1968) 'Lessons from Swedish farm structural policy', *Journal of Agricultural Economics, 19*, 279–99

Williamson, D.F. (1983) 'The workings of the European Commission', *Journal of the Royal Agricultural Society of England, 144*, 23–9

Wilson, G.K. (1977) *Special Interests and Policy Making*, John Wiley, London

6   AGRARIAN REFORM IN EASTERN EUROPE

A.H. Dawson

Agrarian reform is a topic worthy of the geographer's attention. Systems of land tenure show marked spatial discontinuities at national and other administrative boundaries, irrespective of any similarity in the underlying physical environment, and close associations with spatial patterns of land use, agricultural productivity, settlement, trade in farm products, levels of economic development and a wide range of cultural and social characteristics. Changes in those systems can precipitate revolutionary consequences in both the spatial arrangement of the economy and its man-land relationships, and less radical reforms may create hybrid landscapes in which new structures co-exist with earlier forms. Thus, agrarian reform raises issues which are fundamental to a wide range of geographical approaches and topics.

However, most reforms and much of the academic study of them have been concerned with a much narrower range of issues. Firstly, land tenure, and changes within it, have long been seen to be an expression of the real aims and political realities of any society; and secondly, the connection between the system of land holding and the level of development, both in rural areas and in the economy as a whole, has been stressed. Many have been the attempts to reform agrarian structures in order to give expression to newly-adopted social goals or to influence the rate and direction of economic growth, or to achieve both; and there are very few countries in which the state has not felt the need to regulate the system of agricultural land holding at some stage in the recent past, or to alter it from that which had been worked out between landowners and land users, or more often imposed by the landowners, over the course of history.

In these conditions it should not be surprising that the amount of literature about agrarian reform is huge, nor that such terms as 'land reform' and 'structural change in agriculture' should have come to cover a very wide range of interventions by the state. However, it is of some surprise to note that most attention has been paid to the developing countries outwith Europe, to southern Italy and to Japan, in many of which reforms have been of an intermediate 'land-to-the-tillers, compensation-to-the-owners' type (King, 1977; Warriner, 1969), and that study of the much more radical changes in the post-war period in Eastern Europe have been left to the few scholars with particular intersts in centrally-planned economies (Bergmann, 1975; Jackson, 1971; Symons, 1972). It is true that the imbalance of land ownership in Eastern Europe before 1939 was not as great as in Brazil, Chile, Egypt or southern Italy, but the claims which were made for the subsequent collectivisations, and the degree of influence which the state has come to exert upon agriculture, in those countries both mark them out as being of particular interest.

A second reason for the relative neglect of Eastern Europe in recent years may be the impression, at least to the outsider, that agrarian structures there appear to have been rather stable since the period of rapid collectivisation ended in the early 1960s. However, any such impression would be mistaken for not only have the patterns of tenure and farm structure continued to evolve, but also the problems of food supply, which worsened in many countries during the late 1970s, have called in question the whole system of agricultural management and re-awakened the doubts which had been sown by Stalin's earlier collectivisation of Russian agriculture as to whether such a system can ever be successful. However, this account will begin with a brief description of the structure of land tenure as it existed immediately before the Second World War, before turning to, first, the period of rapid change up to 1960, and then the more gradual evolution since then.

The Agricultural Structure of Eastern Europe before the Second World War

Agriculture in Eastern Europe on the eve of the Second World War was characterised by a diversity of structures. The structure in Bulgaria, Romania, much of Yugoslavia, Slovakia an southern Poland (Galicia) was almost entirely made up of small peasant holdings, for land reforms during the inter-war period had contributed to the break-up of estates in all these areas, as had the earlier seizure of lands that had been owned by Turks in Bulgaria, Yugoslavia and Romania. Only 6 per cent of agricultural land in Bulgaria in 1934 was in holdings of 30 hectares or more, and these holdings only represented 1 per cent of the total. Three-quarters of the farms held between two and 30 hectares, but almost one-quarter were dwarf holdings of less than two hectares (Warriner, 1964).

The situation in Albania, Hungary and the eastern provinces of Poland (now in the USSR) was very different. During the 1930s only about 2 per cent of Hungarian farms were of 30 hectares or more, but they included 53 per cent of all agricultural land, and 0.1 per cent occupied more than one-quarter of the land. Similarly, in Albania, in spite of an attempt at land reform in 1930, about one-quarter of the agricultural area was owned by only 3 per cent of the farm households and, in both countries, in consequence, most of the other holdings were small, if not tiny, and incapable of properly supporting those living on them.

The structure of agriculture in the remaining areas of Eastern Europe provided a contrast of another sort. Whereas the areas listed above were dominated either by small or medium-scale peasant farms, on which rather primitive methods of production resulted in low yields, poor standards of living, a large degree of self-sufficiency in the farm economy, considerable debt and an inability to re-invest in the farm unit, or by large estates which employed the landless labourers and owners of dwarf holdings to produce crops for sale, Bohemia, Moravia and western Poland were areas of medium-scale, commercial agriculture. Whereas four-fifths of holdings in the south of Poland in 1931 were of five hectares or less, only about one-third of those in the western part of the country were of that size, and a similar contrast existed between Slovakia and Bohemia in Czechoslovakia. What is more, the farms in western Poland, which had benefited from the development of the economy during the period of Prussian control in the nineteenth century, were better equipped, produced higher yields of crops and milk, and sold a much larger proportion of their products off the farm, than did those in the rest of the country.

Thus, three different farming systems existed in Eastern Europe in the 1930s at the national and regional scales, which reflected the different histories of the area. In some places, such as Romania, extensive reforms had already been undertaken, and land had been parcelled out among peasants, while in others reform had either not been attempted, or had been ineffective, or the scope for it had been very limited. However, in her survey of Eastern European agriculture in the 1930s Warriner (1964) found that land reform had, of itself, had little influence upon the standards of living or the level of economic development. Depressed conditions in world markets had been reflected in low prices for farm goods (especially for those countries in Eastern Europe that found themselves cut off from the German market by tariff walls) and thus had severely inhibited both investment and increased productivity in agriculture, and vitiated the hoped-for effects of land reform, while, in areas in which the scope for reform had been limited and the pressure of rural population heavy, the absence of alternative types of employment either at home or abroad had been reflected in the further subdivision of even some of the smallest holdings, and desperate rural poverty.

The Period of Change 1945–1960

Post-war change may be divided into two periods. Between 1945 and about 1960 rapid and conflicting legislative reforms were enacted, and in some cases rescinded, at a pace which allowed no time for their effects upon the performance of agriculture to emerge in full. Since 1960, in contrast, a longer and more stable — but by no means static — period of tenurial development has occurred, but this has been punctuated by increasingly serious troubles about agricultural supply.

Immediately after the end of hostilities the emphasis in all parts of Eastern Europe was upon reconstruction and the resettlement of lands which had been abandoned by their German, or other, owners. Much of the area that was added to Poland under the Potsdam agreement, and the Sudetenland of Czechoslovakia, had lost their populations, and farmers from other parts of those countries, and in particular from the areas which had been transferred from them to the USSR, were settled on the empty lands.

But land reform was also important, for almost all the newly-established governments in Eastern Europe — which were not initially one-party, communist regimes — were committed to the completion of the land reforms which had begun before 1939. Moreover, they were determined to strengthen those reforms by expropriating the larger holdings without compensation, and by extending that expropriation to middle-sized farms. So, for instance, in Poland all holdings of 100 hectares or more, or which included 50 or more hectares of arable land, were seized, and between 1945 and 1949 almost one-third of the farmland — and four-fifths of that in the former German areas — was redistributed. About 350,000 new farms were created in the area which had belonged to Poland before the war, and about 470,000 in the former German lands. A further one-quarter of a million farms were enlarged (Landau and Tomaszewski, 1985, pp. 187–94).

Similar reforms were also carried out in the other countries. Most of the large estates in Hungary were broken up, and one-third of all agricultural land was redistributed. By 1949, four-fifths of the farmland was in holdings of ten hectares or less, in contrast to two-fifths in 1935 (Dumont, 1957, p. 467). Similar changes, both in extent and effect, were carried out by the Soviet occupation forces in East Germany, and in Czechoslovakia more than one-quarter of the agricultural land was affected (Bergmann, 1975, pp. 79, 102–4). All over Eastern Europe farms of 50 hectares or more in private ownership disappeared, and in some places the maximum permitted size was much less. Everywhere the number of holdings of about five hectares increased. However, in Bulgaria, and to a lesser extent in Yugoslavia and Romania, little land had remained in very large holdings after the inter-war land reforms, and much of that which was distributed had been in the ownership of the churches or Germans (Hamilton, 1968, pp. 171–3).

Thus, considerable changes were made to the pattern of land tenure, and, because the estates were seized without compensation, no financial burden was placed upon the peasants in connection with these changes. (Furthermore, because of other social and institutional changes which occurred during and after the war, the pre-war burden of peasant indebtedness was also removed). However, the reform did not deal effectively with all the inherited problems. For example, many of the farm labourers who received land, and many of those who were resettled, received neither farmhouses nor agricultural buildings, and assistance was required from governments and the United Nations relief agency UNRRA to provide these and minimal levels of working capital. Bergmann notes that many farmers were still lacking such buildings in East Germany in 1953 (p. 103). Secondly, some of those who were resettled found themselves in physical environments of which they had had no previous experience, and in which their methods of farming were inappropriate. What is more, some of the resettled farmers were both conservative and illiterate, and farm output suffered as a result of their slowness in adapting to their new lands (Hamilton, 1968, pp. 171–3). Third, the average size of the new farms was very small, and the total number of very small holdings still remained large, for the underlying pressure of population on the land had scarcely been affected. For instance, in Poland the average size of the new holdings was 6.9 hectares, and the inherited contrasts in structure between the Polish and German areas of the country were not greatly reduced (Table 6.1). On the former German lands the number of holdings of two hectares or less was much reduced by the reform, while the number of those between five and 20 hectares was doubled, but in southern Poland — where so much of the land had been in small peasant farms that little land had been available for parcelling even in the inter-war reform — the proportion of holdings of five hectares or less actually rose from 79.5 per cent in 1939 to 84.7 in 1950. Lastly, the new structure was given far too little time to settle down and reveal any potential which it possessed for economic development for, within two or three years of its creation, attempts were being made by the governments of all the countries of Eastern Europe to alter it fundamentally.

However, before turning to the collectivisation drive of the 1950s,

Table 6.1: Poland - Sizes of Peasant Farms 1939–1950

Area	Size (hectares)	Number of farms (thousands)		Percentage of total	
		1939[a]	1950	1939[a]	1950
Former Polish lands	0–2	783	682	31.3	26
	2–5	782	903	31.2	34.5
	5–10	617	749	24.7	28.6
	10–20	262	251	10.5	9.6
	20–50	58	33	2.3	1.3
	Total	2501	2617		
of which Southern Poland	0–2	236	212	38.1	38.5
	2–5	253	254	41.4	46.2
	5–10	101	76	16.4	13.7
	10–20	23	8	3.6	1.5
	20–50	3	1	0.5	0.1
	Total	618	550		
Former German lands	0–2	144	86	46.6	28
	2–5	45	45	14.4	14.7
	5–10	43	84	13.9	27.3
	10–20	53	76	17.3	24.8
	20–50	24	16	7.8	5.2
	Total	308	307		

Note: a. 1938 for the Former Polish lands.
Source: M. Mieszczankowski (1960) *Struktura Agrana Polski Miedzywojennej,* PWN, Warszawa, pp. 406–13.

it should be noted that not all the land that was seized was subsequently parcelled and given to peasants and landless labourers. Some of the estates, and much of the forest, were retained in the ownership of the state. In some cases, this made good sense in so far as there were insufficient buildings to provide steadings for a large number of independent, peasant holdings in East Germany, the western territories of Poland or in Hungary; and in others estates were used for agricultural training, plant and animal breeding and other specialist activities designed to raise the levels of farm practice and productivity. Nevertheless, a significant part of the land, amounting in the case of Poland to about one-tenth of all the arable and most of the forest — covering about one-quarter of the country — was nationalised.

Collectivisation began in earnest after the establishment of one-party, communist governments in Eastern Europe in the late 1940s. At first, progress was slow, the grouping together of farms largely voluntary, and in some cases the degree of linkage between them rather limited. Members were paid according to the amount of land and other assets which they had contributed. All countries adopted the policy of collectivisation, including Yugoslavia, despite its break with the Soviet Union in 1948. By the late 1950s, in contrast, substantial pressure was being applied to farmers to join, and collectivisation was proceeding rapidly in most countries. In Yugoslavia and Poland, however, the collectivisation drive had collapsed (Table 6.2).

Table 6.2: Land Tenure in Selected Countries in Eastern Europe in 1960

	Percentage of Agricultural Land			Average Size of Farms (ha.)	
	State Farms	Collective Farms	Private Farms	State	Collective
Bulgaria	11.9	86.8	1.3	NA	NA
Czechoslovakia	19.4	65.1	13.5	NA	420
East Germany	6.3	84.4	7.5	604	280
Hungary	13.6	59	27	2,913	935
Romania	11.8	31.5	18.1	3,072	937
Poland	12.7	1.1	85.7	440	128

Source: National Statistical Yearbooks.

Collectivisation of peasant farms was completed in all the other countries by the early 1960s, and only in remote and difficult terrain, such as the moutains of central and eastern Slovakia and of Transylvania, did any substantial area of land remain in independent, peasant farms. The typical collective farm at that time occupied 300–400 hectares, and included between 50 and 100 households. It was usually founded upon a single village. Property divisions in the fields were removed, large areas were sown to a single crop, some collectives built substantial communal barns and byres, and the land was worked by gangs of collective members and their families who were equipped, at least in the early years of collectivisation, from a small number of Machine Tractor Stations (Dumont, 1957, pp. 484–91, 502–10). Most farms operated as single, fully-integrated units, and members were paid, not according to the land or other assets which they had contributed, but for the labour which they provided. Households were allowed to retain small gardens of about half a hectare for their own use. Thus, the structure of agriculture became very similar to that in the Soviet Union, and developed far beyond the many and varied forms of voluntary co-operation which had existed in

Figure 6.1: Agricultural Land in Private Ownership in Poland, 1982

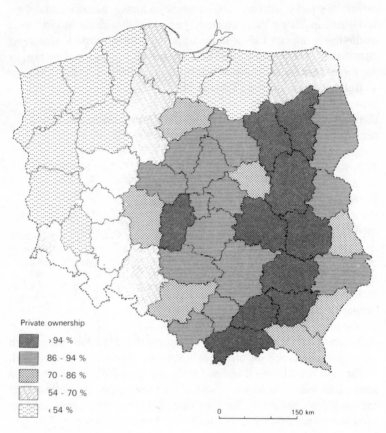

Private ownership

> 94 %

86 - 94 %

70 - 86 %

54 - 70 %

‹ 54 %

0 _____ 150 km

Eastern Europe before 1939, notwithstanding the opposition of many of the farmers to the change.

However, the opponents of collectivisation did prevail in Yugoslavia and Poland. By 1950 almost one-fifth of the agricultural land in Yugoslavia had been collectivised but, under the economic stress that was caused by the rift with the Soviet Union, the government felt obliged to re-assert the right of members to withdraw. By 1953 collectives had fallen to 2 per cent of total agricultural area. The proportion declined still further during the rest of the 1950s, and in 1961 collective farms held no more land than they had in 1946. However, at the same time as the Yugoslav government allowed the break-up of the collectives it lowered still further the maximum area of land which could be held in private to ten hectares, or 15 in areas of poor soil, in an attempt to ensure that capitalist farms could not re-appear. By 1960 only one-tenth of the land was in socialist forms of agriculture — almost entirely in state farms — and peasant holdings of less than five hectares accounted for three-quarters of all the other holdings.

Collectivisation began in Poland in 1949. However, progress was slow, and in 1956 only one-tenth of the farmland had been affected. Furthermore, almost all of this was in the western territories, while the long-settled central and southern parts of the country contained very few collective farms — a pattern which has persisted ever since (Figure 6.1). Following the disturbances of 1956 and the reinstatement of Wladyslaw Gomulka — and opponent of collectivisation — the pressure on farmers to remain within the collectives was removed, and within three months the area of collectivised land had shrunk to less than 2 per cent of the total farmland.

Thus, the period between 1945 and 1960 was one of rapid tenurial change, during which farmers had little time to adapt to any one system, and were frequently assailed by changes in government policy. Until 1953 it appeared as though all the countries of Eastern Europe were progressing in a similar manner, and that the immediate post-war support for the idea of the small peasant farm had gone for good. However, the defection of Yugoslavia and Poland from the collectivist camp has given us some examples by which to assess the success, or otherwise, which might have been achieved elsewhere in agricultural development in the absence of collectivisation, as well as providing us with case studies of the problems that are posed for centrally-planned economies by the presence of a large, privately-owned agricultural sector.

The Period of Consolidation, 1960–1980

By 1980 the agrarian structure of Eastern Europe was different from that of 1960, but the 20-year period had witnessed no reversals of policy comparable with those of 1945 to 1960. Rather, the directions which had been laid out by the late 1950s have been pursued, and the contrasts among the countries which had already been established have survived.

In the case of state farms two major developments occurred in all countries, both of which followed the pattern in the Soviet Union. Firstly, farms became very much larger. For instance, by 1980 the average area of state farms in Hungary had more than doubled, and in Poland it had increased tenfold. In Czechoslovakia it was 6,800 hectares. However, the proportion of the total agricultural area in such farms had not increased to any major extent, for much of the increase in scale had been achieved through the amalgamation of existing units. As a result, the number of farms had fallen to a mere 131 in Hungary, and to 200 in Czechoslovakia, thus allowing much closer control of each by central government, and, perhaps, some economies of scale (Poland, 1982). Secondly, the activities of many state farms had been extended to include some processing of the goods which they produced, and thus they had become a part of the food-processing industry. At the same time, many had also become more specialised, producing a narrow range of related goods on a large scale and in final form, ready for despatch to market. In short, the state sector was more integrated, both horizontally and vertically.

Similarly, there was a general increase in the size of collective farms, together with a fall in their numbers. Those in Czechoslovakia and Hungary had grown to average sizes of about 4,000 hectares, and had come to cover groups of villages, rather than being restricted to the lands traditionally associated with a single rural settlement. However, they had not been linked to the processing industries in the same manner as the state farms. Nevertheless, increased scale was accompanied by more centralised control and, in Czechoslovakia for example, much of the day-to-day decision-making is now in the hands of ministry-approved managers (Cummings, 1982, p. 4). The rights of the members of the collectives were eroded to the point where they are little different from those of the employees of state farms, and the area under private plots on the collectives was reduced.

Pressures were also exerted upon the remaining private farmers during the 1960s and 1970s in most parts of Eastern Europe, with at least three consequences for the agrarian structure. Firstly, in all except Yugoslavia and Poland less than one-tenth of the agricultural land is now in private hands, as collectivisation has been extended to even remote and fragmented holdings in mountainous areas. Secondly, even in Poland the proportion of land in individual farms has been falling, as peasants have been encouraged and cajoled into giving or leasing their farms to the State Land Fund in exchange for a pension, and into leaving the industry. Each year since the late 1950s between 1 and 2 per cent of the agricultural area has been handed over to the Fund — though not all of this has come from private farms — and this land has been distributed, either by sale or lease, to state, collective and private farmers. However, most of the Fund's land has come from the private sector, and most has gone to one or other of the socialist forms of agriculture. For example, between 1978 and 1980, 677,000 hectares were received by the Fund, of which 460,000 came from private holdings, and 832,000 were distributed, but only 334,000 went to individual farmers, and that largely under lease. Thus, the public sector of farming in Poland has been growing slowly, and by 1980 the proportion of farmland in private hands had slipped to about three-quarters. Thirdly, there has been no development in the structure of the surviving private farms. In those countries in which collectivisation of such farms has become almost universal, private farms are not only few, but tiny; and in Poland, where the subdivision of holdings has been forbidden since 1962, it is the stability, not to say rigidity, of their size which has been remarkable in a period in which the scale of enterprise in European agriculture in general has been increasing (Table 6.3). Nevertheless, contrasts in the size of private farms between one part of the country and another are marked, and these reflect not only inherited pre-war characteristics but also the post-war circumstances (Figure 6.2). Thus, while several of the voivodships of the southeast, which had very small holdings before 1939, still have average farm sizes of less than three hectares, and the northeast, with its poor soils and harsh climate, continues to be characterised by farms of twice or three times that size on average, the size of holdings elsewhere in the country has undergone more change. In particular, privately-owned farms in Upper Silesia were subject to much subdivision before 1962 as the rural population was increasingly drawn into industrial employment and became 'worker-peasants', and falls in the areas of holdings around the cities of Warsaw and Krakow occurred for the same reason. Furthermore, the size of private farms in some of the former German lands in the west of Poland has also declined, and in no voivodship does it exceed ten hectares on average.

The Extremes of Eastern Europe in the 1980s

As a result of all the developments during the post-war period in the agrarian structure, some marked contrasts had appeared between the countries of Eastern Europe by 1980, and nowhere were these as great as between Albania, at the one extreme, and Poland, at the other.

Agrarian change in Albania between the 1944 and the mid-1950s followed a similar pattern to that of most of the rest of Eastern Europe, with land reform and then the establishment of state and collective farms,

Table 6.3: Private Farms^a in Poland 1950–1980

	1950	1960	1970	1980
Number (in thousands)	2969	3244	3224	2897
Total area (in thousands of hectares)	17,600	17,400	14,700	13,200
Average size (in hectares)	5.9	5.4	4.6	4.6
Percentage of all farms				
0.5 to 2 hectares	20.9	25.6	26.9	30
2 to 5	33.4	33.7	32	29.5
5 to 10	32.9	28.9	28.5	25.8
more than 10	12.8	11.8	12.6	14.7

Note: a. Of 0.5 hectares or more.
Sources: Poland, *Pocznik Statystyczny 1964,* Warszawa 1964, p. 229;
Rocznik Statystyczny 1983, Warszawa 1983, p. 253.

and machine tractor stations, after the Soviet fashion; and some of the changes since then have also been similar to those in the other countries. For instance, state farms, which cultivated about one-fifth of the sown area in 1979, have been enlarged, and have become specialist enterprises, the rest of the agricultural land is collectivised and the size of collective farms has been increased. Farms in adjacent plain and upland locations have been united in an attempt to reduce the differences between them in the income levels of their members. However, in other ways, the country has followed a different path.

In particular, Albania has remained faithful to Stalin's model of development for a backward farming industry in so far as it has retained machine tractor stations. Whereas other countries disbanded these agencies of central control after Stalin's death, they have been strengthened in Albania, and increased in size, number and range of activities, as a result of which they not only provide collective farms with machinery for the routine tasks of cultivation, but also for land reclamation and such activities as the building of irrigation reservoirs. Furthermore, since 1971 a new type of collective farm has been created — a higher-type collective (HTC) — which enjoys exclusive access to one machine tractor station. By 1981 almost one-quarter of the arable land in the country lay within the HTCs, each of which covered on average about 4,000 hectares, and about one-tenth of the collective farms had been transformed into HTCs.

However, these changes are probably not so much an attempt to strengthen the collective sector in agriculture, as yet another stage in the gradual transformation of collective into state farms. In 1976, following an amendment to the Constitution, collective farmland became state property, and the system of guaranteed payments to workers on state farms has been extended to the members of HTCs. Moreover, the Albanian leader has declared that during the 1981–85 period HTCs should be turned into state farms as they reach comparable levels of efficiency, income and mechanisation (Ballano and Dari, 1984), and that the proportion of the cultivated land in such farms — which stood at about 20 per cent in 1979 — should rise rapidly. It is also envisaged that personal plots on collective farms will disappear for it is argued that they are incompatible with the socialist organisation of production (Madhi, 1982).

Thus, agriculture in Albania will fall increasingly into the direct control of the state, which will not only determine the type and prices of the products, but also the level of income which farmers receive; and the industry will be little different in this or in its greatly increased scale of operations, from manufacturing. Within a period of 40 years, the Albanian authorities have almost entirely supplanted the private ownership of farmland, equipment and livestock — which was deeply entrenched — with Stalinist forms of organisation, and have converted a backward, small-scale, peasant activity into a large-scale, mechanised industry. The range of crops has been increased, and the balance between them improved. Far greater use is now made of irrigation, and much marsh and hill land has been reclaimed (Nuri, 1982). The standard of living in rural areas has been considerably improved (Papjorgji, 1982).

Poland, in contrast, has made relatively modest changes to its agricultural structure since the Second World War. However, by 1980 it was suffering from shortages of farm products and rising imports of fodder crops to such an extent that price rises of 90 per cent were announced by the government, riots and strikes broke out, and the independent trade union, Solidarity, was formed.

The Polish crisis of the early 1980s was caused by a variety of mistakes by the government, but prominent amongst these were errors of agrarian policy. After the dissolution of the collective farms in 1956 more than four-fifths of the farmland was in private hands, but, whereas substantial structural changes (which have been described above) occurred in the state sector, changes in private holdings have been inhibited. During the Gomulka government (1956–1970) investment in private farms was severely constrained by the continuation of the system of compulsory deliveries to the state at low prices, which had existed since the 1940s, and by the exclusion of individual farmers from the national insurance schemes covering illness and old-age pensions, thus obliging them to make provision for these contingencies themselves. Under the Gierek

Figure 6.2: The Size of Privately-owned Farms in Poland, 1982

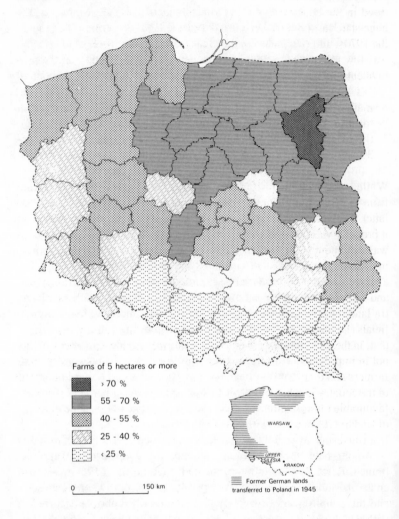

Farms of 5 hectares or more

- ›70 %
- 55 - 70 %
- 40 - 55 %
- 25 - 40 %
- ‹25 %

0   150 km

WARSAW

UPPER SILESIA
KRAKOW

Former German lands
transferred to Poland in 1945

government (1970–1980) these constraints were removed, with the abolition of compulsory purchases and admission to the medical insurance
scheme in 1972, and extension of the pension insurance system to private
farmers in 1977. Moreover, the prices paid by the state for farm products were substantially increased during the 1970s, and free-market

prices rose even more. Thus, farm incomes rose, but agricultural development was still hampered. The farm machinery, that was being manufactured in the country was still of a scale and character more suited to huge state farms; fertiliser became increasingly difficult to buy during the 1970s; and confidence in the long-term future of the private holding was low. Government attitudes towards private agriculture were ambivalent and

> many top managers of the Polish economy remained distrustful towards individual farming and tended to seek methods to accelerate the socialisation of agriculture (Landau and Tomaszewski, 1985, p. 302).

While some well-qualified young farmers were allowed to lease land from the State Land Fund, it became clear that much of the land the latter received or took from farmers who were no longer able to work it properly was destined for the socialist sector, and that the government was still intent in the long run on replacing private with socialised farming. Some enlargement of farms did occur during the 1970s, but the average size of private holdings remained unchanged while their number and total area both fell markedly (Table 6.3), and the acquisition of extra land by private farmers was frequently blocked by government officials (Landau and Tomaszewski, 1985, p. 306). It is hardly surprising that, in these circumstances, much farm income was invested in the 1970s not in improvements to the farms but in very large new houses for the farmers and their families. Furthermore, it was precisely in those areas of the country in which the climatic and pedological conditions are most favourable for agriculture — the southeast — that the highest proportion of land has remained in private ownership, and that diversion of resources from investment was most unfortunate (Dawson, 1982).

Amongst the many demands which Solidarity and its agricultural branch, Rural Solidarity, made during their brief legal existence were an acceleration of the rate of transfer of land between private holdings, and the establishment of a fund, to be financed by foreign donations, with which equipment which was more suitable to the needs of private farms could be imported. Some acceleration did occur in the pace of farm enlargement between 1980 and 1982, but it remains to be seen to what extent the Polish authorities can, or will, allow the appearance of middle-sized, commercial farms of the *kulak* type.

An Assessment

Looking back at the various changes in land tenure in Eastern Europe since the Second World War it could be claimed that the achievements have been great. Agrarian reform has allowed governments to hold down the prices of agricultural products, and thus the level of rural incomes, to divert resources from the farming sector, and to permit the rapid development of other sectors of the economy, and especially the extractive and manufacturing industries. Millions of new, non-agricultural jobs have been created, and in all the countries the proportion of the labour force in agriculture has fallen dramatically (Table 6.4). Moreover, this has occurred despite an increase in the total population in all except East Germany. Only in Albania has the agricultural population risen absolutely, and the effect of this has been offset there by the very substantial programme of land reclamation. Thus, the severe problems of rural population pressure, from which most areas of Eastern Europe suffered before 1939, have been relieved. Landed capital, which was a powerful force in several of the countries, and which would naturally have opposed these policies, has been defeated almost everywhere, and the burden of peasant debt, which bowed the population before the war, and has vitiated many another land reform, has been transformed from an unearned income for former landowners into a source of investment capital for the rest of the economy. In the context of these claims it should be noted that the most serious disturbances to economic growth in Eastern Europe in recent years have occurred in Poland — one of the two countries in which a large private agricultural sector has survived — and have been, in 1970, 1976 and 1980, concerned directly and primarily with the conflict between that sector and the interests of the rest of the community, as that conflict has been expressed in the price of foodstuffs; and it might be concluded that, given that the centrally-planned system of state socialism has been, and will remain, the only likely one in Eastern Europe in the immediate future, there is still scope for agrarian reform in Poland.

However, it would be wrong to discuss the achievements of Eastern Europe in isolation from those of countries which were of similar agrarian structure and performance at the end of the Second World War. Such a comparison is not easy, not least because of the wide variations which existed within some countries at that time and the range of factors, other than land tenure, which affect farm output. Nevertheless, Albania, Bulgaria and Romania might be compared with Turkey and, perhaps, Greece and Portugal; Czechoslovakia with Austria; East Germany with

Table 6.4: Some Indicators of Agricultural Development in Europe since the Second World War

| | Growth of population from 1950 to 1980 % | Percentage of economically active population in agriculture After World War | | 1980 | Yields | | | | | |
		Sec. World War %	Year		Wheat, kg/ha. 1948-52	Wheat, kg/ha. 1980-82	Potatoes, kg/ha. 1948-52	Potatoes, kg/ha. 1980-82	Milk, kg/cow 1948-52	Milk, kg/cow 1980-82
Albania	122	71	1960	60	9.5	24.6	620	750	262	1,735
Bulgaria	22	57	1960	33	12.4	39.9	860	1,050	405	2,718
Czechoslovakia	23	38	1947	10	19.0	42.9	1,130	1,700	1,542	3,190
East Germany	−9	29	1946	10	26.3	45.6	1,610	1,930	2,098	3,894
Hungary	15	53	1949	16	13.8	44.5	680	1,670	1,534	3,605
Poland	44	54	1950	30	12.5	28.7	1,150	1,490	1,770	2,727
Romania	36	74	1950	47	10.2	27.9	790	1,660	896	1,907
Yugoslavia	37	67	1953	37	11.9	32.7	650	920	951	1,653
Austria	8	32	1951	9	17.1	41.6	1,300	2,480	1,836	3,625
Greece	26	48	1951	37	10.2	28.2	1,100	1,800	864	1,947
Italy	22	40	1954	11	15.2	27.0	700	1,820	1,786	2,851
Portugal	18	48	1950	26	7.2	11.3	1;220	870	1,959	2,133
Turkey	118	77	1955	54	10.0	18.7	770	1,660	524	580
West Germany	24	25	1950	4	26.2	51.5	2,110	3,100	2,326	4,537

Sources: FAO *Production Yearbook 1970, 1982*, Rome; United Nations, *Statistical Yearbook 1956*, New York.

West Germany; and Hungary with Italy. If such comparisons are made, Table 6.4 shows that, overall, yields of three major farm products have increased marginally faster in Austria and West Germany than in their Eastern European counterparts, while those in Albania, Bulgaria and Romania have risen by as much as, or more than, in the three countries with which they are linked above, while achieving comparable falls in the dependence of their populations upon farm employment. The performance of Hungarian agriculture, in contrast, appears to have been superior to that of Italy in all the products. In short, there is no clear picture as to which of the two systems of land holding — collectivisation and state farming in Eastern Europe, or the privately-owned farms in the other countries — is the better, and much more detailed comparision is required before a definite answer can be given. However, the detailed results should not be allowed to obscure the fact that in both sets of countries problems of low and stagnant agricultural productivity have obviously been overcome, and that massive improvements have been achieved. Only in Yugoslavia, and, especially, Poland — the country in which the agrarian structure has yet to be stabilised — has agriculture performed less well than in Eastern Europe as a whole.

References

Ballano, P. and Dari, F. (1984) 'The transition to State farming', *Albanian Life, 28,* 15–17

Bergmann, T. (1975) *Farm Policies in Socialist Countries,* Saxon House, Farnborough

Cummings, R. (1982) *A Survey of Czechoslovakia's Agriculture,* United States Department of Agriculture, Washington, DC

Dawson, A.H. (1982) 'An assessment of Poland's agricultural resources, *Geography, 67,* 297–309

Dumont, R. (1957) *Types of Rural Economy,* Methuen, London

Hamilton, F.E.I. (1968) *Yugoslavia: Pattern of Economic Activity,* Bell, London

Jackson, W.A.D. (1971) *Agrarian Policies and Problems in Communist and Non-Communist Countries,* University of Washington Press, Seattle

King, L. (1977) *Land Reform,* Bell, London

Landau, Z. and Tomaszewski, J. (1985) *The Polish Economy in the Twentieth Century,* Croom Helm, London

Madhi, R. (1982) 'The process of strengthening the socialist psychology of property and work', *Albania Today, 63,* 25–33

Nuri, F. (1982) 'Achievements in land reclamation and irrigation', *Albania Today, 66,* 43–5

Papjorgji, H. (1982) 'Peopling the countryside and extending the working class to the whole territory of Albania', *Albania Today, 63,* 14–19

Poland (1982) *Kraje RWPG 1982,* Glowny Urzad Statystyczny, Warszawa

Symons, L.J. (1972) *Russian Agriculture: a Geographic Survey,* Bell, London

Warriner, D. (1964) *Economics of Peasant Farming,* Frank Cass, London

Warriner, D. (1969) *Land Reform in Principle and Practice,* Clarendon Press, Oxford

7 AGRICULTURE AND URBAN DEVELOPMENT
C.R. Bryant

Introduction

The theme of agriculture and urban development — the conversion of agricultural land to urban uses, the mixing of agricultural and urban land uses and the ramifications of these processes — has probably been the most frequently studied component of the broader thrust of research on urbanisation-agriculture interactions. It has spawned extensive research and widespread concern over the implications of the conversion and mixing processes for agricultural production systems in many developed countries, including the US (e.g. Fletcher and Little, 1982; Hart, 1976; the NALS studies, 1980, 1981), the UK (e.g. Best, 1974, 1976, 1977, 1981; Centre for Agricultural Strategy, 1976; Coleman, 1978a), New Zealand (e.g. Coleman, 1967) and Canada (e.g. Bryant and Russwurm, 1979; Gierman, 1977; Krueger, 1959, 1978; McCuaig and Manning, 1982).

While the focus of this chapter is on agriculture and urban development, a concern primarily relating to the land demand emanating from urban areas, it is important to stress the scope of the broader urbanisation-agriculture interaction thrust. This broader theme, a major research area in agricultural geography, also has strong links with several other sub-fields of geography such as urban geography and resources management as well as with a number of other disciplines and professions such as land use planning, agricultural economics, land economics and rural sociology. Inevitably then, some of the scholarly work cited in this chapter has been produced by non-geographers; however, it has important geographic implications or components and it has certainly influenced the work by agricultural geographers. The widespread interest in urbanisation-agriculture interaction from many disciplines and sub-disciplines in itself hints at the complexity of the processes under investigation, and suggests that researchers and scholars focusing on the agriculture-urban development theme should be careful to tease out the interrelationships and implications of the broader range of urbanisation-agriculture interaction processes for their own specific research problem.

Indeed, the broad scholarly and professional interest in agriculture and urban development may be partly responsible for the increasing recognition that has been given since the early 1970s to the complexity of the processes affecting agriculture around cities throughout the Western world (e.g. Bryant, 1976, 1981; Bryant and Greaves, 1978; Bryant *et al.*, 1982; Moran, 1979; Munton, 1974a; Smit and Joseph, 1982). First, explicit recognition has been given to the complexity created by the variety of stimulating, constraining and permissive forces impinging upon agriculture and how they combine (e.g. Best, 1981; Bryant, 1976, 1981, 1984a; McCuaig and Manning, 1982; Munton, 1974a). At the most general level, it has been argued that urbanisation, the process by which an increasing proportion of the population of a country lives in 'urban areas and zones', generates a threefold set of demands which may elicit a response from, and thus have an impact upon, agriculture (Bryant, 1976, 1984a; Bryant *et al.*, 1984). This threefold set of demands comprises a demand for *land* for various urban-related uses and functions, a demand for *labour* resulting from expanison in non-farm employment opportunities and a demand for *agricultural produce and services* associated with both the spatial concentration of demand in 'urban' zones and the increased standards of living that have often been linked to urbanisation in the Western world. Specifically, then, a set of *urbanisation* forces other than just the demand for land for residential, industrial and infrastructural development has been identified and their role in modifying agriculture-urban development interaction processes stressed. Furthermore, other forces and influences unrelated to urbanisation or metropolitan processes have also been recognised, e.g. technological change in agriculture and competition from other regions or countries (e.g. Bryant, 1976; Munton, 1974a), and the role of variations between regions and subregions in terms of the physical, economic and cultural environment highlighted (e.g. Best, 1981; Bryant, 1984b; Moran, 1979). It is suggested that the ways in which these various *urbanisation* and *non-urbanisation* forces combine to produce different net positive or negative effects upon agricultural productivity and structure may be of immense significance in influencing what we are able to observe about the agriculture-urban development interaction process and, indeed, how we evaluate it.

Second, attention has been drawn to the problems and complexities related to the geographic scale at which the various processes linking agriculture with urbanisation and urban development specifically, operate. Macro-scale differences between metropolitan regions in terms of the nature and, presumably, the relative importance of the underlying forces and processes have been identified in several countries, including the UK (e.g. Best, 1981; Champion, 1974; Thompson, 1981), Canada (e.g. Bryant, 1976; Gierman, 1977) and the US (e.g. Ziemetz *et al.*, 1976).

Perhaps more important, given a concern for understanding process, attention has been drawn to variations between farm entrepreneurs because of the importance of the individual farm entrepreneur in evaluating the resources, opportunities and constraints facing the farm (Olmstead, 1970) and in the subsequent choice of strategies leading to farm change (e.g. Bryant, 1973, 1981; Moran, 1979). Thus, a variety of farm-level factors that influence, condition, constrain or stimulate farm decisions must also be taken into account at certain scales of analysis. Such factors can be seen as directly related to the individual farm itself (e.g. farm size, specific field layout, 'inherited' enterprise structure) or to the farmer and his/her farmily (e.g. age structure, potential farm family continuity, non-farm investment portfolio).

Nonetheless, while these complexities have undoubtedly become increasingly recognised, geographic work on the agriculture-urban development theme is still dominated by studies and approaches in which narrower perspectives have been taken. It will be argued that, while significant contributions to our understanding have been made by these studies, it is partly in response to their shortcomings that the broader perspectives have begun to emerge. These broader perspectives remain, however, largely at the level of conceptualisations and interpretative devices; they are as important in laying out directions for future research in this field as they are for any inherent explanation they provide in themselves.

A number of literature reviews dealing with the general topic or segments of it have already appeared (e.g. Bryant and Russwurm, 1979; Bryant *et al.*, 1982; Furuseth and Pierce, 1982a; Munton, 1974a; Pacione, 1984; Wibberley, 1959). Therefore, the thrust of this chapter is more conceptual, especially in terms of dealing with the more recent developments, and methodological. In order to provide a statement on progress in agriculture-urban development research, the first task is to define the scope of the field and to identify the significant sets of questions in it. Second, the major preoccupations of researchers in this field are set against this backdrop. In discussing the degree to which the open questions in the field have been answered, a number of interpretative and methodological biases are discussed. Finally, in conclusion, directions for future research are laid out.

Agriculture and Urban Development: Scope

The scope of research in this field can be defined most easily in relation

Figure 7.1: The Agriculture–Urban Development Interaction Process

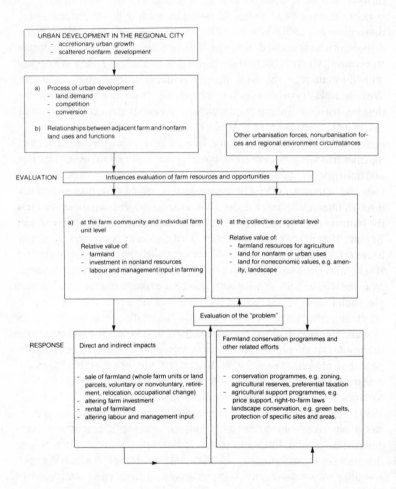

to the broad process by which urban development interacts with agriculture (Figure 7.1). This can be reduced essentially to the impact that urban development has on agriculture. Any effect that agricultural structure might have on the growth patterns of urban land uses could be seen more properly as a concern for urban geography, although the manner in which non-farm land uses and functions disperse around cities may be influenced by certain aspects of agricultural structure which in turn, of course, may have some impact on the ongoing agricultural system. The term 'urban development' is used to refer both to accretionary growth at the edges of cities and to the development of more dispersed non-farm land uses and functions around cities (Bryant and Russworm, 1979). This broad view of urban development is in keeping with the nature of the predominant evolving settlement form in the Western world in the post-industrial age, viz. the *regional city*. This settlement form is characterised by a built-up core, relatively easily associated with 'urban', and its surroundings made up of a mix of open countryside, villages, towns and dispersed settlement. The whole is tied together into a functioning entity by movements of people, goods, money and information; it can be regarded as urban because various urban land uses and activities are spread throughout it yet are linked to each other through interaction into a distinctive settlement form. That part beyond the built-up core has been referred to as the *urban field* (Russwurm and Bryant, 1984), a more restrictive use of the term compared to earlier usage (cf. Friedmann, 1973; Friedmann and Miller, 1965; Hodge, 1974). If several regional cities merge, then the settlement system moves away from the regional city form towards a megalopolitan structure, with multiple nuclei, intervening countryside and complex interaction patterns (Gottman, 1961).

Land demand and 'consumption' for urban land uses and functions *at the edge* of the built-up core of the regional city and *within its urban field* thus provide, the *raison d'être* of agriculture-urban development research. Therefore, the effects of urban areas as concentrated markets for agricultural produce (cf. AREEAR, 1976; Gregor, 1963; Laureau, 1983; SEGESA, 1973), as generators of opportunities for enterprises based on excess farm resources and as concentrations of non-farm employment opportunities (cf. Bryant, 1980; Pautard, 1965) are not, in themselves, of central concern in this chapter. However, together with non-urbanisation forces they must be considered as modifying influences of some significance in the agriculture-urban development interaction process (cf. Moran, 1979). Similarly, another urbanisation-related process influencing labour, viz. the reverse movement of some city people back into farming to create part-time farms and hobby farms, is not of central concern in itself for studying agriculture-urban development yet its results may be significant if it means that people with different motivations for owning farmland are introduced into an area since their responses to potential land demand for urban land uses and functions may be quite different from those of the previous farmland owners (cf. McKay, 1976).

Logically, two broad sets of questions appear to be of significance in research into the agriculture-urban development interaction process (Figure 7.1). First, there are questions concerning the *process* by which urban development has an impact on agriculture. There are two subsets of impacts, related to different stages of the urban development process (Figure 7.2): (a) those impacts that are created by the *actual process of development* itself, viz. land conversion and land market effects; and (b) those impacts relating to the *final destination uses and functions* and their relationships with the ongoing and adjacent agricultural structure. Important research questions can be classed into three groups: (a) the identification of the various impacts and their extent; (b) the evaluation of these impacts from the perspective of the farming community and individual farm; and (c) understanding how the responses to the development pressures are translated into changing agricultural structures and land use (Figures 7.1 and 7.2). In considering the last two sets of research questions, it will be argued that it is particularly important to view the farm unit and the agricultural structure from a systems perspective (Bryant, 1984b; Olmstead, 1970).

The second broad set of questions concerns the collective, or societal, evaluation of, and response to, the impacts alluded to above. Important research questions comprise the *evaluation of the impacts* of urban development on agriculture from the *collective* perspective, the *evaluation of alternative forms of intervention* and the *role and effect of different forms of intervention* in the agriculture-urban development interaction process.

Research has been undertaken on a wide range of topics related to these broad sets of research questions. However, as is argued below, progress in our understanding has been very uneven and there have been some noticeable biases in the dominant research thrusts.

Agriculture and Urban Development: Preoccupations and Achievements

The Process

In relation to the impacts of urban development on agriculture, it is useful to distinguish between the *triggers* or changes in the farmer's environment created by actual and potential urban development and the decisions

Figure 7.2: Impacts of Urban Development on the Individual Farm Unit

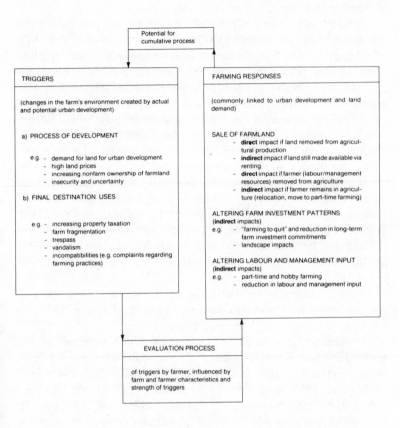

made by farmers in response to these triggers after an evaluation process has been undertaken (Figure 7.2). Both are clearly part of 'impact', yet they have not been consistently differentiated in the literature and there have been few attempts to link farmers' responses to specific triggers or combinations of triggers.

An important distinction has, however, been made between *direct* and *indirect* impacts of urban development on agriculture (Bryant and Russwurm, 1979; Pacione, 1984) (Figure 7.2). Direct impacts refer to the removal of agricultural resources from production while indirect impacts refer to effects imparted to the ongoing agricultural structure. It would appear reasonable to reserve the terms direct and indirect impacts to refer to farmers' responses and therefore to the resulting adjustments to farming and farm structure.

Direct impacts in terms of the land resource occur primarily at the penultimate and final stages of the land conversion process; they may occur earlier if idling of the land resource accompanies, for instance, extensive land speculation. For the most part, this removal of land from agricultural production can be seen as negative from the perspective of the maintenance of a viable agricultural production system on the land, but, of course, individual farmers and landowners may view it as personally very positive! Direct impacts on labour resources involve farmers making decisions leading to retirement or complete occupational change.

Indirect impacts occur within the ongoing agricultural structure in terms of changes in farm structure, its viability and productivity. Frequently assumed to be negative, it has been suggested that the net effect of these impacts is collectively of greater significance than the direct impacts of removal of land from agriculture and actual land conversion (cf. Krueger, 1959; Rodd, 1976), even though there are some potentially positive effects such as increased farmland rental opportunities. In addition, recognition must be made of the indirect impacts on the *agricultural landscape* because of the significance that landscape and amenity values hold in many countries such as the UK (e.g. Munton, 1983a,b) and France (e.g. Bryant, 1984a).

The research literature abounds with examples of statements on the negative effects of urban development upon agriculture, particularly from North America (e.g. Hoffman, 1982; Krueger 1959, 1978, 1984; Raup, 1975; Rodd, 1976; Russwurm, 1977; Sinclair, 1967) and the UK (e.g. Coleman, 1978a; Standing Conference, 1976; UK Ministry of Agriculture, Fisheries and Food, 1977), but also from other countries such as Italy (e.g. Sermonti, 1968) and France (e.g. Biancale, 1982). The overwhelming impression from much of the literature is still that agriculture around cities is fighting a losing battle, that concrete and urban people will overrun the land and lay it to waste and that nothing short of public intervention on a significant scale can stop the inevitable destruction of viable agriculture there.

Closer investigation reveals, however, considerable ambiguity and debate, even in terms of identifying the extent of direct impacts. Research on the direct impacts has focused naturally on land conversion or consumption rates, especially the conversion of prime quality agricultural land both in North America (e.g. Crerar, 1961; Gierman, 1977; Krueger, 1978; Peterson and Yampolsky, 1975; Vining *et al.,* 1977) and in

Western Europe (e.g. Best, 1979, 1981; Coleman, 1978b). There have been long-standing debates on both sides of the Atlantic both in terms of the interpretation of the evidence used to measure land conversion rates and of the evaluation or significance of the rates — a debate more properly related to the second broad set of research questions concerning the agriculture-urban development interaction process. Much of the debate regarding interpretation of evidence concerns the nature of the data sources used. Two sources of information on land conversion, or more generally land use change, have been commonly used, viz. official statistics of the census variety and direct observation of land use change through air photo analysis and/or land use mapping exercises.

The use of official statistics has focused on agricultural census material. Best (1981) provides an excellent discussion of the advantages and pitfalls of using this type of material in the British context, but much of his discussion is equally applicable to other countries (see, e.g., Bryant *et al.*, 1982). First, decreases in the area of farmland recorded in census counts have often been used in discussions of farmland losses. Frequently, there has been an explicit link made between negative changes or losses in farmland area and urban growth; the classic examples are where land losses have been related to a measure of population growth, e.g. the indices of the area of farmland losses per 1,000 increase in population in Bogue's (1956) and Crerar's (1961) studies. The interpretative problem is that a change in census farmland recorded between two years signifies simply, and only, that the land managed within census farms or holdings has changed; where a loss has occurred, the data generally do not indicate to what degree the loss is a result of real land use conversion to urban uses or to other uses (e.g. forestry), land abandonment or 'accounting transfers' related to changing definitions of census farms. Despite the difficulties of interpreting such data, they continue to be used in discussions of farmland losses to urban development. Even in the UK, where the annual agricultural returns made by farmers include information on the 'change in the area of holding' to indicate the broad intended uses to which agricultural land has been transferred, Best (1981) laments the continued misinterpretation of these data. *Total* farmland reduction is still frequently attributed to urban development instead of focusing on transfers specifically to urban development — and this is quite apart from problems of variable availability of the data over time, changes in definitions and plain errors.

The use of direct observation of land use transfers has become more popular since the mid-1970s particularly through air photo analysis. Large-scale exercises based largely on field observation and mapping, such as the Land Utilisation Surveys in the UK, have encountered difficulties associated with timeliness and temporal comparability because field programmes have usually been spread over several years. Even with some of the exercises involving air photos, lack of uniform photo coverage at comparable points in time has necessitated extrapolation which is open to question when patterns of development can vary so dramatically over relatively short periods (cf. Gierman, 1977). All in all, while our potential ability to measure land use change has undoubtedly increased substantially, problems related to resources, timeliness and comparability continue to create circumstances in which the interpretation of the evidence is open to debate (see Bryant and Russwurm (1983) for a review of attempts at monitoring land use change in North America and Western Europe).

In terms of indirect impacts (Figure 7.2), there has been less *specific* research even though these impacts are widely recognised in the research literature. One of the most significant areas that has developed is concerned with the effect that *expectations* of urban growth might have upon agricultural investment and the associated atmosphere of *uncertainty* in the vanguard of rapid urban development, a phenomenon which has strong ties to land speculation. Although the germs of the ideas can be traced to the 1950s (e.g. Wibberley, 1959), the first significant formalisation is found in Sinclair's now classic paper (1967) which has had a significant influence on much subsequent geographic work on agriculture around cities (e.g. Boal, 1970; Bryant, 1974, 1981; Mattingly, 1972). The ideas developed by Sinclair (1967) and refined by Bryant (1974) were articulated as a *partial model* linking urban development, the shortening of farm investment planning horizons and farm investment under a set of circumstances that included rapid and continuous urban development (explicitly stated) and a reasonably good physical environment for agricultural production (implicit). Put simply, it was hypothesised that, where expectations of urban development were high, the planning horizons for farm investment would be shortened thus discouraging longer term investments and even leading to disinvestment (e.g. cutting down on maintenance and renewal of tile drainage systems).

The mechanisms implied are intuitively appealing and have been widely referred to in the literature. However, the actual evidence of the extent and significance of this indirect impact is scanty and ambiguous. There is some evidence of a reduction in certain indicators of intensity of farm production in some metropolitan regions (e.g. Thompson, 1981) but whether this can be ascribed to the uncertainty or anticipation of the urban development mechanism or whether it simply reflects other broader processes at work in the agricultural industry is unclear. Some work which claims to provide evidence of the Sinclarian pattern of agricultural intensity near cities is less clear upon closer examination. Mattingly (1972), for instance, uses census-based measures of labour intensity per acre and per farm on a township basis for an area in north central Illinois to show lower patterns of intensity in 'urban' townships. However, the specific relationship regarding uncertainty and agricultural intensity really necessitates data on certain types of capital investment (Sinclair (1967) hints at this, and Bryant (1974) develops this consideration explicitly). Moreover, there are a host of other characteristics such as part-time farming and farm-type (recognised by Mattingly) that are related to production intensity and which can be influenced by forces other than the uncertainty consideration. As another example, in a detailed farm-level investigation (Bryant, 1981), the relationship remained ambiguous with no clear line being identified between a lowering of the intensity of farm investment and the strength of the indicators of potential urban development. The reasons for this ambiguity are discussed below. Furthermore, the period of rapid growth has turned out to be short-lived in many areas, and downward revisions to urban growth estimates have been commonplace — in the Paris region, for instance, population projections made in the mid-1960s for the region were around 14 million for the year 2000 AD, but in 1980 proposed revisions to the regional master plan population forecasts for the year 2010 AD were placed at closer to 11 million (IAURIF 1980).

Generally, the literature dealing with the indirect impacts of urban development upon agriculture has attempted either: (a) to draw inferences about those impacts based upon macro-analysis of surrogate variables; (b) to focus on one specific type of impact; or (c) simply to enumerate the variety of impacts. Good examples of the first type are found in attempts to test Sinclair's hypothesis (e.g. Mattingly, 1972) and also in statements and analyses concerning the effect of high land prices and urban growth. It has frequently been assumed, for example, that high land prices create difficulties for farmers who wish to expand their land base to remain competitive and that production costs are adversely affected. Assuming farmers prefer to purchase additional land, the observation of high land prices is then assumed to indicate a 'problem' (e.g. Morris, 1978; Rodd, 1976). However, detailed evidence does not always support such inferences, partly because of the increased popularity of farmland rental as a means of farm business expansion especially in North America (cf. Bryant and Fielding, 1980; Greaves, 1984) and partly because of the existence of other counteracting forces.

The best example of work focused on to one specific type of impact is probably the body of literature dealing with the impact of urban-related development upon agricultural property taxation. The notion is that where farm real estate is taxed as a source of municipal revenue farmers in the areas surrounding cities may pay more in property taxes than their non-farm neighbours or farmers in other areas because either the valuation of farm real estate is affected by urban influences and/or because municipal costs are greater with the additional demands created by the non-farm population. The literature is overwhelmingly North American (e.g. Krueger, 1957; Plaut, 1977; Roberts and Brown, 1980; Walrath, 1957); and it is more extensively developed in the area of modifications of systems of property evaluation and taxation than in terms of the evaluation of the significance of increased taxes to the farmer, though there are some notable exceptions (e.g. Plaut, 1977). Research that has tackled the significance of this phenomenon has not found evidence to support the hypothesis that higher real property taxes lead to an increase in the loss of farmland or that it forces farmers to sell land to speculators or developers (e.g. Plaut, 1977). Whether or not this factor influences farming practices, however, has still not been adequately tackled.

Finally, statements on many of the other indirect impacts are characterised by being enumerative and by a focus on the presence of the *trigger* (Figure 7.2) rather than an evaluation of the significance of the impacts. It has become commonplace to list the various triggers such as trespass, vandalism (including uncontrolled garbage disposal), farm fragmentation and general incompatibilities between agriculture and non-farm uses (for recent surveys of such impacts, see Bryant *et al.*, (1982) and Pacione (1984)). It is rare to see any attempt to evaluate the real significance and extent of these or, indeed, of any of the indirect impacts; some of the work referred to earlier regarding higher property taxes for farmers near cities provides notable exceptions.

In summary, research on the process by which urban development creates impacts on agriculture has been maintained at a high level of activity for over two decades. Significant advances have been made (see below), yet a stereotyped image of agriculture being on a downhill path around cities still prevails. Why is this so? It is suggested that a number of interpretative biases and methodological features of much of the research provide a partial explanation for this.

Perhaps the most significant interpretative bias is the frequently implied assumption that agriculture near cities is responding to metropolitan forces, primarily urban development-related, *and* that the impact of these forces is essentially negative for agriculture. This downplays the role of other urbanisation or metropolitan forces as well as non-urbanisation influences (Figure 7.1) and the positive effects of urbanisation forces such as market effects and increased opportunities to rent farmland. These other influences are important in two respects. First, they may modify significantly the net impact on agricultural structure of the negative urban development forces, even to the extent of yielding an overall positive net impact on agriculture. Opportunities to short-circuit marketing channels may permit farmers to develop 'pick-your-own' enterprises (e.g. Laureau, 1983; Moran, 1979), farm shops and other enterprises based on renting out 'excess' farm resources, e.g. use of obsolete barns for storage of recreational vehicles and boats and the boarding of horses. Furthermore, increased non-farm ownership of farmland may provide real opportunities for farm expansion of a relatively permanent nature; here, the direction of the impact depends to a great extent upon the conditions surrounding the lease *and* how it is perceived by the farmer. Where farmland has been bought by non-farm interests either as an incidental purchase along with a farmstead for a weekend or seasonal retreat or as a hedge against inflation in an area remote from any development potential, the leasing arrangement can give rise to long-term improvement in the farm structure. Where the non-farm interest is concerned with speculation and/or land development, the chances are increased that the leasing situation will be viewed as precarious and that agricultural disinvestment might ensue. However, even here such arrangements may last considerable lengths of time, especially where an active period of land speculation is followed by a general slowing down of economic growth (Bryant and Fielding, 1980; Munton, 1982, 1983b).

Second, from the perspective of evaluating the significance of the impacts for the farm operation, other processes may be influencing the agricultural structure and creating a more important set of problems to which the farm entrepreneur must respond. Depending on the extent to which this exists, it can create difficulties in establishing any relationships specifically between agricultural change and urban development, and it provides a partial explanation of some of the ambiguities in the evidence.

Much of the research has also been characterised by a number of methodological thrusts. First, many studies have relied almost exclusively on *secondary data sources,* particularly agricultural census data and aerial photography. The work dealing with agricultural land removal has been mentioned earlier, but, in addition, census-type data have been used to investigate the changing agricultural structure around cities, using a variety of scales of geographic analysis, both in the UK (e.g. Thompson,

1981) and in North America (e.g. Bryant, 1976; INRS, 1973; McCuaig and Manning, 1982). Important contributions have undoubtedly been made by such studies, particularly in relation to identifying patterns that suggest the existence and importance of processes other than urban development forces. The most useful contributions appear to be those where either a comparative component was an explicit part of the research design or where the range of geographic situations covered was large enough for a comparative component to evolve naturally.

Studies of agricultural land conversion rates have been vastly improved upon through the use of more problem-specific secondary data. Use of aerial photography to measure land conversion more directly has resulted in a more accurate assessment of land conversion rates (e.g. Gierman, 1977) and broad geographic coverage has given fuel to the arguments that the processes influencing agricultural land reduction vary tremendously from region to region. North American studies have led the way in the use of aerial photography to investigate land conversion rates around cities at a national scale — see especially the Environment Canada studies in Canada (Gierman, 1977; Warren and Rump, 1981) and the 1976 US Department of Agriculture study (Ziemetz *et al.*, 1976). In the latter study, substantial regional differences were identified in the relationship between urban development and change in the agricultural land base; some urbanising counties, for instance, experienced overall increases in cropland while others experienced cropland reduction following transfer of land to forests or simple land abandonment. The study concluded that changing agricultural technology and productivity in other regions was the cause of much of the cropland reduction.

For England and Wales, use of the more problem-specific 'change in area' data from the annual agricultural returns (Best, 1981; Best and Champion, 1970) provides further evidence of substantial regional variations in the transfer of agricultural land to urban uses, variations that are not related in any simple way to population growth. Even the recent use of the less problem-specific census farmland change variable has yielded more useful results when researchers have approached the problem forewarned of the complexities of farmland reduction. Bryant *et al.*, (1981) were able to demonstrate how 'out-of-line' earlier estimates of agricultural decline had been (Crerar, 1961) simply by performing a geographic disaggregation of the earlier data and highlighting the substantial variation in farmland reduction rates within a region that could not be attributed to urban growth differentials. And Crewson and Reeds (1982) in an analysis of farmland loss based on agricultural census data for an area on the edge of the Toronto urban field in Ontario were able to demonstrate a complex set of factors unrelated to non-farm development. Nevertheless, Best (1981) warns of continuing misinterpretations of even the more problem-specific data, and Allison (1984) was still able to conclude in an analysis of the Golden Horseshoe counties in Southern Ontario that the loss of 30 to 47 per cent of their farmland area from 1941 to 1976 was almost solely because of urban development forces. The existence of considerable areas of technologically marginal farmland in this otherwise quite heavily urbanised region was thus ignored.

Research using census data to study changing agricultural structure near cities has been able to suggest the complexity of processes affecting agriculture there. In an extensive analysis of Canadian agriculture around the major cities, strong regional differences in terms of farmland reduction, farm amalgamation and farmland tenure changes have been highlighted, differences that are partly related to differences in the broad regional environment for agriculture as well as to differences in the urbanisation environment (Bryant, 1976). Furthermore, agricultural changes in urban-centred regions have been shown to bear strong similarities to adjacent hinterland (or non-urban) regions, emphasising the importance of processes tied to the characteristics of the broad regional environment (Bryant and Russwurm, 1981; Bryant *et al.*, 1984). Yet, despite these advances, there are serious limitations to how much further our understanding of the processes can go using such data. Where research incorporates correlation, regression and factor analytic approaches, scale of analysis and interpretation problems abound. Finally, we are frequently left to deal with surrogate variables because there are constraints in these data bases that militate against detecting directly certain significant positive changes, e.g. direct selling, or how farmers might respond to some pressures via disinvestment. In the final analysis, there can be no substitute for problem-specific data collected at the farm level.

A second methodological thrust, which is also partly an interpretative bias, has been a strong emphasis on the *land* resource. This is partly related to the frequent use of secondary data sources that do not permit farm-level analysis, even though many different scales of spatial unit have been used in data collection and analysis. The emphasis on land (cf. the agricultural *land* problem, agricultural *land* conservation) has led to the people and capital components of agricultural production often being downplayed.

Even where farm-level data are acquired, it has usually been easier to identify the *existence* of potential impacts whose trigger has an obvious physical manifestation on the land (e.g. trespass, vandalism, farm fragmentation) or that otherwise have been given a high profile (e.g. higher land prices and property taxes) than to assess the real significance of these impacts on the farm. Phrased in the terminology of Figure 7.2, it is easier to get farmers to acknowledge the existence of the triggers than to get them to evaluate their importance and to assess their farming response. Researchers have thus encountered difficulty in both defining and collecting problem-specific data *and* in placing what evidence they do have on impacts in a broader context. One notable exception is Munton's (1983b) work using farm-level data on London's Metropolitan Green Belt in which he was able to test a number of hypotheses linking agricultural land maintenance to various factors such as short-term letting and proximity to development and highways. This work, however, emphasises the landscape maintenance aspects of the management of agricultural land and does not deal very directly with agricultural productivity relationships.

It is argued that the *farm* has all too often not been treated as a *unit* in analyses, and that specific relationships have been investigated without asking *how important* they are in the context of the overall farm structure and changes therein. Even though the measurement of the significance of farmers' responses to the various triggers may be extremely difficult, treating the farm as a whole unit, and thus taking a systems perspective, does provide a partial solution. Specifically, farmers can be questioned about the existence of various triggers (or stimuli) over a period of time and about any changes that have been made on the farm over the same period of time; analysis would then be attempted of the relationships between the farmers' responses on triggers and changes. It is too easy simply to ask farmers for their experiences with certain triggers or development pressures (e.g. 'Check whether you have encountered difficulties with any of the following . . .') and then, on the basis of positive responses, to assume that agriculture is going downhill. The difficulties encountered with various development pressures may be real, but they may not always be significant in relation to difficulties with origins elsewhere or to the opportunities present. For example, in an analysis of a farm sample from the Paris region, Bryant (1981) found that farmers' evaluations of a non-agricultural future for their land and the range of problems encountered related to urban development pressures were indeed correlated with a set of measures of the strength of urban development pressures and potential. However, neither evaluations nor problems encountered nor strength of the actual urbanisation presence were found to be linked to actual patterns of farm investment change. The explanation advanced was that although farmers in areas with high development potential recognised the development pressures and associated problems, many were also able to take advantage of the opportunities presented by a near-urban location, e.g. direct sales of produce to the consumer and 'pick-your-own' enterprises. Furthermore, many of the development-related problems recognised, e.g. trespass and vandalism, created more of a nuisance value than major financial problems, a conclusion supported by research elsewhere (e.g. UK Ministry of Agriculture, Fisheries and Food, 1977).

This general situation has been exacerbated on occasion by a third methodological feature, viz. the fact that many studies have dealt with very small geographic areas, almost entirely located within the urban fringe (cf. Krueger's work (1959, 1978, 1984) in the Niagara Fruit Belt in Canada). The danger arises of 'sampling-at-the-margin', and specifically of not being able to separate out what Martin (1975) has called problems 'of' the urban fringe from problems 'in' the urban fringe ('in' implying that they occur in the urban fringe *and* elsewhere). This is particularly dangerous if reliance is placed on secondary data sources (official statistics, air photos) or field observation without a questionnaire survey. Broader geographic coverage or a comparative component to the research design allows this problem to be tackled.

Despite the above discussion, the evidence has been growing that the frequently espoused stereotyped image of agricultural responses to urban development pressures near cities hides a much more complex reality. Important regional and subregional differences have been highlighted (e.g. Best, 1981; Best and Champion, 1970; Bryant, 1976; Ziemetz *et al.*, 1976) and it is increasingly clear that farmers do not react in the same way to the same development pressures or triggers (Figure 7.2) (e.g. Bryant, 1981; Moran, 1979; Munton, 1983b). Some farmers, for instance, evaluate positively some of the impact triggers such as rising farmland values and greater non-farm ownership of farmland (due to greater farmland rental opportunites) (e.g. Bryant and Fielding, 1980; McRae, 1981; Smit and Flaherty, 1980). Furthermore, at certain levels non-farm development can benefit agricultural communities through the infusion of capital and population to create a more viable community (e.g. Dahms, 1980), and even the long-held belief that farmers and non-farmers have markedly different service demands has been subject to question (e.g. Joseph and Smit, 1981; Smit and Joseph, 1982). Perhaps most important of all, healthy farm communities exist near cities in some areas where the conventional wisdom would place agriculture on the path to elimination (e.g. Berry, 1978; Bryant, 1981; Punter, 1976).

In response to the shortcomings and the emerging evidence, more comprehensive conceptual frameworks have been developed as noted earlier. Recently, the author (Bryant, 1984a) has suggested that the urbanisation and non-urbanisation forces, and regional environment influences, combine to produce different types of farming landscape change, and a threefold classification was suggested. Both agricultural productivity and, to a lesser extent, landscape issues are addressed in this framework, positive and negative impacts recognised and the importance of entrepreneurial adaptive behaviour stressed. First, in *landscapes of agricultural degeneration*, negative forces are dominant. For instance, negative urban development impacts may combine (e.g. high land prices, excessive farm fragmentation, uncertainty and vandalism) to thwart agricultural progress, overriding other potentially beneficial urbanisation forces (e.g. market access), regional environment circumstances (e.g. good quality land resources, favourable market conditions for the dominant enterprises in a region) and non-urbanisation forces (e.g. import controls on competing produce). It was suggested that often negative non-urbanisation and regional environment circumstances are also present in landscapes of agricultural degeneration. Thus, there is no attempt to deny the negative impacts of many urban development factors, but rather the suggestion is that agricultural degeneration results more generally from a *combination* of such factors, frequently interacting with other negative factors unrelated to urban development.

Second, *landscapes of agricultural adaptation* involve some of the negative impacts being present, but being outweighed by positive urban development effects (e.g. farmland rental opportunities), other positive urbanisation forces (e.g. market access) and non-urbanisation and regional environment circumstances. Farmers are able to adapt to the negative influences and take advantage of ongoing opportunities so that the agricultural structure 'progresses'. Thus, just because there are triggers present associated with potentially negative impacts does not mean that the net result of the interplay of the whole range of factors impinging upon agriculture must be negative. Finally, where urban development-related forces are minimal, this second type merges into *landscapes of agricultural development*. Here, the area's evolution is a function of the interplay of non-urbanisation forces, regional environment circumstances and certain urbanisation forces (e.g. labour competition from the non farm sector), although negative changes in non-urbanisation forces could force adaptation or even initiate degeneration. In both landscapes of agricultural adaptation and agricultural development, the potential for complementary relationships between urbanisation forces in the broadest sense and agricultural change is recognised.

This framework is stressed as an interpretative device and conceptual framework. The important point in the argument is not that the pressures from urban development are not real, but that there are circumstances where either they may not be significant relative to other pressures or that farmers may be able to adapt to or to neutralise them because of other positive influences in their environment. None of the forces acts in isolation, so to understand the relative importance of negative urban development impacts one must attempt to comprehend the total set of forces affecting agriculture near cities.

Societal Response

Research involving the need for, nature of and role of public intervention in the agriculture-urban development interaction process has often resulted in informative, yet descriptive, accounts of various programmes and strategies. Much of the work has been undertaken by non-geographers although geographers have made some important contributions. There is an important role for the descriptive studies, especially in disseminating information regarding different approaches and what forms of intervention are available. More important questions for research in the long term include the evaluation of the 'problem' and goals, the evaluation of the internal consistency of programmes and their external consistency with respect to stated goals and the evaluation of the results of public intervention.

As with the literature dealing with 'process', the 'problem' has generally been assumed to be *real* and *significant* without critical evaluation. The 'So what?' question has so often not been asked critically (Best, 1981; Frankena and Scheffman, 1980; Johnston and Smit, forthcoming). This is partly related to the state-of-the-art in research dealing with process and partly to the strong focus on land again — it can be argued that if there is an agricultural land resource problem it is only a symptom of a broader people and capital problem. Yet this perspective has received, comparatively, little attention (but see, e.g. Lapping and Fitz-Simmons, 1982). It is not being argued that there are no problems, but rather that the evidence is not subjected sufficiently to critical evaluation.

Programme statements themselves often contain questionable overstatements regarding the extent of the 'problem' (e.g. OMAF, 1977; Québec, 1979). Best (1981) in summing up his analysis of land use change suggests:

land use planning in Britain — and in most other countries as well — has been built upon the extremely shaky and insecure foundation of illusion rather than that of reality. (Best, 1981, p. 184)

In particular, he shows that urban sprawl has not been progressing at the 'alarming' rates claimed by many conservationists and that the argument that agricultural output and food supplies (in Great Britain) will be seriously endangered by continuing urban encroachement on farmland does not withstand analysis. This does not mean that careful planning should not continue to husband our agricultural resources, but it does mean that some of the reasoning put forward for widespread intervention may be on shaky ground. Part of the dilemma facing conservationists has been the massive shifts in agricultural productivity and serious over-production problems in many Western countries during the past three decades. Munton (1983a) notes that the value of domestic agricultural production, including food exports, was equal to, in 1980,

74.8 per cent of indigenous-type food consumed in the United Kingdom and 60.5 per cent of all food . . . (while) . . . comparable figures for 1969–71 were 58.6 per cent and 45.9 per cent respectively. (Munton, 1983a, p. 360)

The important societal question then is not so much the quantity of prime agricultural land being removed for urban uses but how the amount relates either to the total supply of such land and, more especially, the capacity of the food supply system to replace it from other sources. Indeed, one must ask whether it is even significant to attempt to replace such 'lost' capacity.

In the research literature on the societal response to agriculture-urban development interactions, there has been relatively little attention paid to the role of variations in the 'regional environment', particularly in terms of differences in the cultural and political dimensions of different environments. Yet this dimension may influence how the problem is perceived relative to other issues, how intervention is perceived and, given some intervention is deemed necessary, what a 'reasonable' form of intervention might be in a given context. Part of the problem is the lack of comparative frameworks or research in which it is recognised that goals, for instance, may vary from place to place and time to time. A good example of the importance of this is the much greater value attached to the landscape amenity functions of farmland in Western Europe (e.g. France) compared to North America (Bryant, 1984c). Some work has involved international comparisons (e.g. Bryant and Russwurm, 1982; Furuseth and Pierce, 1982b) but these are rare. One result of the lack of comparative analysis is the tendency to argue that the specific situation between agriculture and urban development in a given country is unique. Canada, so the conventional wisdom would hold, is unique because despite its large land area, the area of high quality agricultural land is very small and coincident with the largest urban areas; similarly, Britain is unique in terms of the degree of land use competition it has experienced because of its small size and high level of urbanisation. Such statements must be treated with caution. Best (1981) uses a careful, and cautious, analysis of land use data from the EEC countries and North America to show several recurring features and particularly that the UK situation is not all that remarkable.

The overall socio-economic, demographic and political context in which farmland and agricultural issues exist may change over time. Priorities attached to different issues may therefore change. Frequently, discussions in the professional geographic literature on agricultural land conservation implicitly take on the position that nothing can be more significant than the protection of agricultural land from urban develop-ment in an agricultural area, regardless of changing context.

However, quite apart from the relative importance of the need for housing, infrastructure and industry, other conflicts in the countryside have attracted more and more attention (e.g. Davidson and Wibberley, 1977). Examples are the conflict between modern farming practices and the long-term agricultural productivity of the land because of links to soil erosion and agricultural pollution (e.g. the Sparrow Report, 1984). While there is ongoing debate regarding the gravity of the conflicts, the latter relationship may provide a stronger rationale for conserving prime agricultural land than arguments based on conversion rates of agricultural land to urban uses alone. The amenity concerns have been particularly important in the UK and some other West European countries. In North America, a complex mix of situations exists, but generally amenity values and long-term agricultural productivity issues arising from the nature of modern farming practices around cities have only recently begun to emerge as significant issues for public debate.

Research that has attempted to evaluate the operation of various pro-grammes has had to grapple with developing appropriate measures of 'success' (Bryant and Russwurm, 1982). Most of the work has been undertaken by non-geographers though there are some notable excep-tions (e.g. Pierce, 1981; Wilson and Pierce, 1984). What are the ap-propriate yardsticks? As soon as a sophisticated approach to land issues is accepted, i.e. where all land uses have a potential claim on the land even where some form of preferred zones for agriculture are identified, or where we simply accept that a number of goals must be weighed simultaneously, it becomes more and more difficult to determine what constitutes 'success' and what not.

A final comment concerns the relative lack of attention given to the farm business perspective. Lapping and FitzSimmons (1982) stress how important farm viability issues are in dealing with farmland conservation. Perhaps of even greater importance is our lack of understanding of how farm entrepreneurs respond and react to different forms of public intervention. This is an issue of considerable import if, indeed, as has been suggested elsewhere (cf. Bryant *et al.*, 1982), 'persuasive-regulatory' approaches to management based upon co-operation and persuasion are becoming more important.

Conclusions

A number of future research directions appear to follow logically from this appraisal of existing work in the agriculture-urban development field and how it has addressed the important questions. In terms of research into the process by which urban development has an impact on agriculture, perhaps the greatest need is for considerably more attention to data collection and analysis *at the individual farm business level.* This would help improve our state of knowledge about the relative importance of different types of impact as well as allow the field to advance towards theory-building beyond the current conceptual frameworks and interpretative devices. This thrust would also help identify the relative importance of urban development-related impacts compared to other types of change in farm structure and farming practices. It is suggested that an approach using a systems perspective is worth considering; this would involve a focus on the *whole* farm unit by identifying specific time segments and paths along which a farm evolves at different periods in its life, and linking each to the generic circumstances surrounding each segment. In the necessary farm-based surveys, it will be important not to load questions and allow *a priori* assumptions about the relative importance of different circumstances to creep in. Finally, sorting out the relative importance of different processes and forces will be greatly facilitated by a greater attention to the careful design of a comparative component in such work.

The need for more comparative work is also apparent in research dealing with the societal response to the agriculture-urban development process. In analysing the societal response, it is imperative to see the 'problem', the response and its sufficiency in the context of the other processes affecting agriculture and in the context of other legitimate concerns requiring the allocation of public resources. Finally, the role of the individual farm entrepreneur — how he/she responds to different programmes, what role he/she has to play in them — deserves much more attention than has been the case so far.

In conclusion, the agriculture-urban development field has attracted an enormous amount of attention in the geographic literature and related fields since the late 1960s. The sheer volume in the English language alone is impressive. Significant advances have been made in documenting and understanding specific parts of this complex interaction process. Yet there is much left to be accomplished and a recurring theme in this chapter has been the argument that further significant advances in this field will depend upon the acquisition and analysis of problem-specific data at the farm level, all placed within a comparative and more comprehensive framework. Happy are those researchers whose 'problem' can be treated independently of other processes — but such is not the lot of students of the agriculture-urban development interaction process.

Acknowledgements

The initial preparation and writing for this chapter was undertaken during the 83–84 academic year while the author was on sabbatical leave from the University of Waterloo, supported by a Leave Fellowship from the Social Sciences and Humanities Research Council of Canada which is gratefully acknowledged. The development of the ideas expressed owes much to discussions that I have had over the last few years with several colleagues, especially Lorne Russwurm of the University of Waterloo and Richard Munton of University College, London — to them I extend my thanks for all the time they have given me. Finally, I would like to thank Tom Johnston, a PhD candidate at the University of Waterloo, for his critical commentary on an earlier draft of the chapter and for his time spent in discussions.

References

Allison, R. (1984) *Agricultural Land Use and Farming Systems in Southern Ontario*, College of St. Mark and St. John, Canadian Studies Geography Project for Sixth Forms and Colleges, Plymouth, England

ARREAR (Atelier Régional d'Etudes Economiques et d'Aménagement Rural) (1976) *L'Agriculture Spécialisée en Ile de France*, Ministry of Agriculture, Paris, France

Berry, D. (1978) 'Effect of urbanization on agricultural activities', *Growth and Change*, *9*, 2–8

Best, R.H. (1974) 'Building on farmland', *New Society* 31 Oct., 287–8

Best, R.H. (1976) 'The extent and growth of urban land', *'The Planner'* (Journal of the Royal Town Planning Institute.) *62*, 8–11

Best, R.H. (1977) 'Agricultural land loss — myth or reality?' *The Planner* (Journal of the Royal Town Planning Institute.) *63*, 15–16

Best, R.H. (1979) 'Land-use structure and change in the EEC', *Town Planning Review*, *50* (4), 395–411

Best, R.H. (1981) *Land Use and Living Space*, Methuen, London

Best, R.H. and Champion, A.G. (1970) 'Regional conversions of agricultural land to urban use in England and Wales', 1945–67, *Transactions of the Institute of British Geographers*, *49*, 15–32

Biancale, M. (1982) *Les Coteaux de Chambourcy à Orgeval*, Institut d'Aménagement et d'Urbanisme de la Région d'Ile-de-France, Paris, France

Boal, F.W. (1970) 'Urban growth and land value patterns', *Professional Geographer*, *22*, 79–82

Bogue, D.J. (1956) *Metropolitan Growth and the Conversion of Land to Non-Agricultural Uses*, Scripps Foundation Series in Population Distribution, Oxford, Ohio

Bryant, C.R. (1973) L'agriculture face à la croissance métropolitaine: le cas des exploitations de grande culture expropriées par l'emprise de l'Aéroport Paris-Nord', *Economie Rurale*, *95*, 23–5

Bryant, C.R. (1974) 'The anticipation of urban expansion: some implications for agricultural land use practices and land use zoning', *Geographia Polonica*, *28*, 93–115

Bryant, C.R. (1976) *Farm-Generated Determinants of Land Use Change in the Rural-Urban Fringe in Canada, 1961–1975*, Technical Report, Lands Directorate, Environment Canada, Ottawa

Bryant, C.R. (1980) 'Manufacturing in rural development' in D.F. Walker (ed.) *Planning Industrial Development*, John Wiley and Sons, Chichester, England, Ch 5. pp. 99–128

Bryant, C.R. (1981) 'Agriculture in an urbanizing environment: a case study from the Paris region, 1968 to 1976', *Canadian Geographer*, *21* (1), 27–45

Bryant, C.R. (1984a) 'The recent evolution of farming landscapes in urban-centred regions'. *Landscape Planning*, *11*, (4), 307–26

Bryant, C.R. (1984b) 'Agriculture in the urban fringe: a systems perspective', *Rural Systems*, *2*,

Bryant, C.R. (1984c) 'Farmland conservation and farming landscapes in urban-centred regions: the case of the Ile-de-France region', manuscript, Department of Geography, University of Waterloo, Canada

Bryant, C.R. and Fielding, J.A. (1980) 'Agricultural change and farmland rental in an urbanising environment', *Cahiers de Géographie de Québec*, *24*, 277–98

Bryant, C.R. and Greaves, S.M. (1978) 'The importance of regional variation in the analysis of the urbanisation-agriculture interactions', Cahiers de Géographie de Québec, *22*, 329–48

Bryant, C.R. and Russwurm, L.H. (1979) 'The impact of non-agricultural development on agriculture', *Plan Canada*, *19* (2), 122–39

Bryant, C.R. and Russwurm, L.H. (1981) 'Agriculture in the urban field, Canada, 1941 to 1971' in K.B. Beesley and L.H. Russwurm (eds.), *The Rural-Urban Fringe: Canadian Perspectives*, Geographical Monograph 10, Department of Geography, Atkinson College, York University, Toronto, pp. 34–52

Bryant, C.R. and Russwurm, L.H. (1982) 'North American farmland protection strategies in retrospect'. *Geojournal*, *6* (6), 501–11

Bryant, C.R. and Russwurm, L.H. (1983) *Area Sampling Strategies in Relation to Land Use Monitoring Needs and Objectives*, Working Paper 24, Lands Directorate, Environment Canada, Ottawa

Bryant, C.R., Russwurm, L.H and McLellan, A.G. (1982) *The City's Countryside: Land and its Management in the Rural-Urban Fringe*, Longman, London

Bryant, C.R., Russwurm, L.H. and Wong, S.Y. (1981) 'Census farmland change in Canadian urban fields, 1941–1976.' *Ontario Geographer, 18*, 7–23

Bryant, C.R., Russwurm, L.H. and Wong, S.Y. (1984) 'Agriculture in the Canadian urban field: an appreciation' in M.F. Bunce, and M.J. Troughton, (eds.), *The Pressures of Change in Rural Canada*, Geographical Monograph 14, Department of Geography, Atkinson College, York University, Toronto, Ch.2, pp. 12–33

Centre for Agricultural Strategy (1976) *Land for Agriculture*, Report 1, University of Reading, Reading, England

Champion, A.G. (1974) *An Estimate of the Changing Extent and Distribution of Urban Land in England and Wales, 1950–70*, RP 10, Centre for Environmental Studies, London

Coleman, A.M. (1978a) 'Planning and land use', *Chartered Surveyor, 111* (5), 158–63

Coleman, A.M. (1978b) 'Agricultural land losses: the evidence from maps' in A.W. Rogers (ed.), *Urban Growth, Farmland Losses and Planning*, Wye College for the Institute of British Geographers, England

Coleman, B.P. (1967) 'The effect of urbanisation on agriculture' in J.S. Whitelaw (ed.), *Auckland in Ferment*, New Zealand Geographical Society, Auckland, New Zealand, pp.102–111

Crerar, A.D. (1961) 'The loss of farmland in the growth of the metropolitan regions of Canada' in *Resources for Tomorrow: Supplementary Volume*, The Queen's Printer, Ottawa, pp. 181–96

Crewson, D.M. and Reeds, L.G. (1982) 'Loss of farmland in south-central Ontario from 1951 to 1971', *Canadian Geographer, 26* (4), 355–60

Dahms, F.A. (1980) 'The evolving spatial organization of settlements in the countryside — an Ontario example', *Tijdschrift voor Economische en Sociale Geografie, 71*, 295–306

Davidson, J. and Wibberley, G.P. (1977) *Planning and the Rural Environment*, Pergamon, Oxford, England

Fletcher, W.W. and Little, C.E. (1982) *The American Cropland Crises*, American Land Forum, Bethesda, Maryland

Frankena, M.W. and Scheffman, D.T. (1980) *Economic Analysis of Provincial Land Use Policies in Ontario*, Ontario Economic Council, Toronto

Friedmann, J. (1973) 'The future of the urban habitat' in D.M. McAllister (ed.), *Environment: a New Focus for Land-Use Planning*, National Science Foundation, Washington

Friedmann, J. and Miller, J. (1965) 'The urban field', *Journal of the Institute of American Planners, 31*, 312–20

Furuseth, O.J. and Pierce, J.T. (1982a) *Agricultural Land in an Urban Society*, Association of American Geographers, Resource Publication in Geography, Washington, DC

Furuseth, O.J. and Pierce, J.T. (1982b) 'A comparative analysis of farmland preservation programmes in North America', *Canadian Geographer, 26*, 191–206

Gierman, D.M. (1977) *Rural to Urban Land Conversion*, Occasional Paper 16, Lands Directorate, Environment Canada, Ottawa

Gottman, J. (1961) *Megalopolis: the Urbanized Northeastern Seaboard of the United States*, The Twentieth Century Fund, New York

Greaves, S.M. (1984) *Farmland Rental and Farm Enlargement: a Southern Ontario Example*, unpublished PhD thesis, Department of Geography, University of Waterloo, Waterloo, Canada

Gregor, H.F. (1963) 'Industrialised drylot farming: an overview', *Economic Geography, 39*, 299–318

Hart, J.F. (1976) 'Urban encroachment on rural areas', *Geographical Review, 66*, 1–17

Hodge, G. (1974) 'The city in the periphery' in L.S. Bourne (ed.), *Urban Futures for*

Central Canada: Perspectives on Forecasting Urban Growth and Form, University of Toronto Press, Department of Geography Publications, Toronto, pp. 281–300

Hoffman, D.W. (1982) 'Saving farmland, a Canadian program', *Geojournal, 6,* 539–46

IAURIF (Institut d'Aménagement et d'Urbanisme de la Région d'Ile-de-France) (1980) *Projet de Schéma Directeur d'Aménagement dt d'Urbanisme de la Région d'Ile-de-France*

INRS (Institut National de la Recherche Scientifique) (1973) *Région Sud: l'Agriculture*, Office de Planification et de Développement du Québec, Montreal

Joseph, A. and Smit, B. (1981) 'Implications of exurban residential development: a review', *Canadian Journal of Regional Science, 4,* 207–24

Johnston, T. and Smit, B. (forthcoming) An evaluation of the rationale for farmland preservation policy in Ontario. *Land Use Policy*

Krueger, R.R. (1957) 'The rural-urban fringe taxation problem: a case study of Louth Township', *Land Economics, 33,* 264–9

Krueger, R.R. (1959) 'Changing land use patterns in the Niagara fruit belt', *Transactions of the Royal Canadian Institute, 32* (67) (Part 2), 39–140

Krueger, R.R. (1978) 'Urbanization of the Niagara fruit belt', *Canadian Geographer, 22,* 179–94

Krueger, R.R. (1984) 'The struggle to preserve speciality crop land in the rural-urban fringe of the Niagara peninsula of Ontario' in M.F. Bunce and M.J. Troughton (eds.), *The Pressures of Change in Rural Canada*, Geographical Monograph 14, Department of Geography, Atkinson College, York University, Toronto, Ch. 15, pp. 292–313

Lapping, M.B. and FitzSimmons, J.F. (1982) 'Beyond the land issue: farm viability strategies', *Geojournal, 6,* 519–24

Laureau, X. (1983) 'Agriculture péri-urbaine: des enterprises pour demain', L'Agriculture d'Enterprise (Bulletin of the Centre de l'Agriculture d'Enterprise, Paris) 171–2, 3–42

Martin, L.R.G. (1975) *A Comparative Urban Fringe Methodology*, Occasional Paper 6, Lands Directorate, Environment Canada, Ottawa

Mattingly, P.F. (1972) 'Intensity of agricultural land use near cities', *Professional Geographer, 24* (1), 7–10

McCuaig, J.D. and Manning, E.W. (1982) *Agricultural Land Use Change in Canada: Process and Consequences*, Land Use in Canada Series 21, Lands Directorate, Environment Canada, Ottawa

McKay, R.D. (1976) *The Land Use Characteristics and Implications of Hobby Farming: a Case Study in the Town of Caledon*, unpublished BES thesis, School of Urban and Regional Planning, University of Waterloo, Waterloo, Canada

McRae, J.D. (1981) *The Impact of Exurbanite Settlement in Rural Areas: a Case Study in the Ottawa-Montreal Axis*, Working Paper 22, Lands Directorate, Environment Canada, Ottawa

Moran, W. (1979) 'Spatial patterns of agriculture on the urban periphery: the Auckland case', *Tijdschrift voor Economische en Sociale Geografie, 70,* 164–76

Morris, D.E. (1978) 'Farmland values and urbanization', *Agricultural and Economic Research, 30* (1), 44–7

Munton, R.J.C. (1974a) 'Farming on the urban fringe' in J.H. Johnson (ed.), *Suburban Growth: Geographical Processes at the Edge of the Western City*, John Wiley and Sons, London, pp. 201–23

Munton, R.J.C. (1974b) 'Agriculture and conservation in lowland Britain' in A. Warren and F.B. Goldsmith (eds.) *Conservation in Perspective*, John Wiley and Sons, Chichester, England, pp. 323–36

Munton, R.J.C. (1982) 'Land speculation and the underuse of urban fringe farmland in the Metropolitan Green Belt', Paper presented at the Anglo-Dutch Symposium on *Living Conditions in Peri-urban and Remote Rural Areas in North West Europe*, University of East Anglia, September 1982

Munton, R.J.C. (1983a) 'Agriculture and conservation: what room for compromise? in

A. Warren and F.B. Goldsmith (eds.) *Conservation in Perspective,* John Wiley and Sons, Chichester and London, Ch. 20, pp. 353–73

Munton, R.J.C. (1983b) *London's Green Belt: Containment in Practice,* John Wiley and Sons, Chichester and London

NALS (National Agricultural Lands Study) (1981) *Final Report,* US Government Printing Office, Washington, DC

Olmstead, C.W. (1979) 'The phenomena, functioning units and systems of agriculture', *Geographica Polonica, 19,* 31–42

OMAF (Ontario Ministry of Agriculture and Food) (1977) *Green Paper for Planning for Agriculture: Foodland Guidelines,* The Queen's Printer, Toronto

Pacione, M. (1984) *Rural Geography,* Harper and Row, London

Pautard, J. (1965) *Les Disparités Régionales dans la Croissance de l'Agriculture Française,* Série Espace Economique, Gauthier-Villars, Paris

Petersen, G.E. and Yampolsky, H. (1975) *Urban Development and the Protection of Metropolitan Farmland,* Urban Institute, Washington

Pierce, J.T. (1981) 'The BC Agricultural Land Commission: review and evaluation', *Plan Canada, 21* (2), 48–56

Plaut, T. (1977) *The Real Property Tax, Differential Assessment and the Loss of Farmland on the Rural-Urban Fringe,* Discussion Paper Series 97, Regional Science Research Institute, Philadelphia

Punter, J.V. (1976) *The Impact of Exurban Development on Land and Landscape in the Toronto-centred Region, 1954–1971,* Central Mortgage and Housing Corporation, Ottawa.

Québec (1979) *Loi sur la Protection du Territoire Agricole: Renseignements Généraux,* Commission de Protection du Territoire Agricole du Québec, Québec

Raup, P.M. (1975) 'Urban threats to rural lands: background and beginnings', *Journal of the American Insititute of Planning, 41,* 371–8

Roberts, N.A. and Brown, H.J. (1980) *Property Tax Preference for Agricultural Land,* Allanheld, Osmun and Co. and Lincoln Institute of Land Policy, Montclair, New Jersey

Rodd, R.S. (1976) 'The crisis of land in the Ontario countryside', *Plan Canada, 16,* 160–70

Russwurm, L.H. (1977) *The Surroundings of Our Cities,* Community Planning Press, Ottawa

Russwurm, L.H. and Bryant, C.R. (1984) 'Changing population distribution and rural-urban relationships in Canadian urban fields, 1941–1976' in M.F. Bunce and M.J. Troughton (eds.) *The Pressures of Change in Rural Canada,* Geographical Monograph 14, Department of Geography, Atkinson College, York University, Toronto

SEGESA (Société d'Etudes Géographiques, Economiques et Sociologiques Appliquées) (1973) *L'Agriculture Spécialisée de la Région Parisienne Face à la Croissance de l'Agglomération,* Report to the Ministry of Agriculture, Paris, France

Sermonti, E. (1968) 'Agriculture in areas of urban expansion: an Italian study', *Journal of the Town Planning Institute, 54,* 15–17

Sinclair, R.J. (1967) 'Von Thünen and urban sprawl', *Annals of the Association of American Geographers, 57,* 72–87

Smit, B. and Flaherty, M.F. (1980) 'Preferences for rural land severance policies: an empirical analysis', *Canadian Geographer, 24,* 165–76

Smit, B. and Joseph, A.E. (1982) 'Trade-off analysis of preferences for public services', *Environment and Behaviour, 14,* 238–58

Sparrow Report (1984) *Soil at Risk: Canada's Eroding Future,* Report on Soil Conservation by the Standing Committee on Agriculture, Fisheries and Forestry to the Senate of Canada, Ottawa

Standing Conference (1976) *The Improvement of London's Green Belt,* Standing Conference on London and South-east Regional Planning, London

Thompson, K.J. (1981) *Farming in the Fringe,* CCP 142, Countryside Commission, Cheltenham, England

U.K. Ministry of Agriculture, Fisheries and Food (1977) *Peri-urban Agriculture in the*

Slough-Hillingdon Area (Region of London), Report presented to Conference on *Peri-urban Agriculture*, OECD, Paris

Vining, D.R. Jr., Plaut, T. and Bieri, K. (1977) 'Urban encroachment on prime agricultural land in the United States', *International Regional Science Review*, 2, 143–56

Walrath, A.J. (1957) 'Equalization of property taxes in an urban-rural area', *Land Economics, 33* (1), 47–54

Warren, L. and Rump, P. (1981) *Urbanization in Canada, 1966–1976*, Land Use in Canada Series 20, Lands Directorate, Environment Canada, Ottawa

Wibberley, G.P. (1959) *Agriculture and Urban Growth: a Study of the Competition for Rural Land*, Michael Joseph, London

Wilson, J.W. and Pierce, J.T. (1984) 'The Agricultural Land Commission of British Columbia' in M.F. Bunce and M.J. Troughton (eds.) *The Pressures of Change in Rural Canada*, Geographical Monograph 14, Department of Geography, Atkinson College, York University, Toronto, Ch. 14, pp. 272–91

Ziemetz, K.A., Dillon, E., Hardy, E.E. and Otte, R.C. (1976) *Dynamics of Land Use in Fast Growth Areas*, Agricultural Economics Report 325, Economic Research Service, US Department of Agriculture, Washington

8 LAND OWNERSHIP AND THE AGRICULTURAL LAND MARKET

P.J. Byrne

Introduction

Financial institutions, in particular the pension funds and insurance companies, influence life in Britain in a wide variety of ways. One of the principal ways in which this influence is felt, both directly and indirectly, is through their massive ownership of land and buildings, and their continuing large scale involvement in the markets for these commodities. It is of course the medium to long term financial return on these investments which has attracted and held the interest of these institutions and which in turn has regularly drawn political and media responses, usually critical, to the activities of the institutions in all sectors of the property market — residential, commercial and agricultural.

The question of ownership by such institutions in the agricultural sector was raised most recently in the Parliamentary debates on the Agricultural Holdings Bill (Hansard, 1983, 1984). The extent to which they are responsible for taking tenanted land out of the market was questioned and the fear voiced that they were replacing that form of tenure with in-hand, managed, farming — 'agribusiness'. The view was also expressed that even where they retained tenants, it was without any real commitment in the form of further investment to the enterprise itself.

The last time these same questions were raised was in 1977, when the then government set up a committee of enquiry to report on the state of agricultural land acquisition and occupancy. The committee reported in May 1979, (Northfield, 1979) and presented a fairly definitive view of the subject at that time, extending its view beyond the rather limiting question of institutional ownership to give wider consideration to trends in acquisition and occupancy, as its terms of reference required.

The interest of financial institutions in agricultural land has been a major influence on the market during the last decade which has in any case been a period of very rapid evolution in British agriculture. It is the extent to which the effects of that involvement can be observed and interpreted which is the main theme of this chapter.

Agricultural Land Market Data and their Limitations

Statistics on land ownership and occupancy are sparse and we were hampered in our work because of this. There is a need for much more information to be made available as a base for policy decisions affecting agricultural structure and in the long term, a full system of land registration with land use identified and beneficial interests visible, would be desirable. (Northfield, 1979)

These remarks serve to identify what continues to be by far the most severe problem in discussing land ownership in Britain. This is the fact that reliable and consistent information is not publicly available. It is ironic therefore that it is really only because the financial institutions, as opposed to individuals, have been slightly more open about their investments and investment policies that it has been possible to gather any real evidence on which to base expressions of concern as to what their influence might be on the long term structure of agriculture in Britain. Even so, the relative availability of land ownership information on these institutions as a group is very much a special case. Although it may be possible to identify the person or company presently occupying a piece of land, there is really no public way of establishing the actual ownership of that land in any but a very small fraction of cases. There can be good commercial reasons why, at the time of sale, the price paid for a piece of land may need to be kept confidential. As will be shown, price data are not the main problem, ownership is. Even if the disinclination to be open about ownership were to be overcome, a data collection system for a full scale cadastral survey would be so expensive to establish that it now seems unlikely ever to happen. It can be done, however, and by comparison other countries have good cadastral databases either in existence or under development which identify the spatial and legal characteristics of land and property in a comprehensive way and, more importantly, make these data available at reasonable cost to any party (see for example Collins and Smith, 1979 and Kalms, 1985). There was strong support for such a database in the Northfield report and this interest and support has continued up to the present (RICS, 1977). Under the terms of the Agriculture (Amendment) Act 1984 it is now legally possible for government to collect information on agricultural occupancy, and so, for this section of the property market at least, some improvement in published data may be expected. As to what is already possible, in Scotland, ownership can be identified in many cases through the Register of Sasines, although the process is labourious, given the fragmented structure of ownership. In contrast, information held by the Land Registry in England and Wales is incomplete and makes no distinction between legal ownership and occupation. Access is restricted to competent users and the information is arranged in ways which make statistical analysis so difficult as to be pointless. Complete information on ownership, held for taxation purposes by the Inland Revenue, is held in absolute confidence, under the Official Secrets Act, and is unlikely ever to be made available in a disaggregated form.

Types of Agricultural Land Ownership

The distinction between occupation and ownership is significant. In England and Wales all land is technically and legally owned by the Crown. As such it is formally a right or interest which is 'purchased' when property is exchanged. A single piece of land may have numerous interests associated with it at any one time, and this adds to the complexity of determining ownership. Ownership in this sense means the possession of the freehold interest, that is the land is held from the Crown without any dues being paid and with the allowance of inheritance at the freeholder's will. This tenurial right usually has attached the legal 'Estate' of Fee Simple Absolute. This gives to the freeholder the right to lease part of their 'Estate' to a lesser interest if they wish for some term of years. The lessee may then in turn sublet part of his interest for a period of years, although that period must be less than that of the next higher lease and may be subject to covenants restricting the extent of subletting. Clearly the potential result is a confusing mass of interests in a single piece of land, making actual ownership virtually impossible to establish if the owner so desires, and it seems that most do!

Thus the land tenure system has basically two categories, freehold ownership and tenancy, although within each category there are a number of subgroupings and arrangements. Within the owner-occupied group, for example, it is possible to identify a number of subdivisions: farmed in hand with a manager; through a partnership; via a farming company; or farmed by the owner or his family directly.

Tenants of let farms have gained a considerable security of tenure under successive Acts of Parliament, a position which seems unlikely to be affected to any great extent by the provisions of the new Agricultural Holdings Act 1984. The main provision of this Act is the abolition of statutory succession on the death of a tenant whose tenancy was granted on or after 12 July 1984. Tenancies granted before that date still have the right of succession for up to three generations, and it will clearly be many years before any except marginal effects on the tenure system become apparent. A particular exception to this could well be the case, more frequent of late, where an owner occupier sells his farm then leases it back, usually on a modern full tenancy agreement, with until recently the security of three generations of tenure. Sale and leaseback could well become a much less popular selling option, although there must remain some circumstances in which the exchange of security for capital will be perceived as appropriate for an individual. Equally the buyer in such cases, often a financial institution, is not then likely to wish to offer an extended tenancy clause in any leaseback arrangement since it would clearly restrict future investment decisions and reduce the resale value of the property. This is because one of the most obvious effects of the protection that tenants have enjoyed has been the development of what is effectively a submarket for tenanted land. The freehold of such land will sell for less than equivalent vacant possession land, reflecting the encumbrance of a tenant. This difference is known as the vacant possession premium.

There has been a steady decline over the last 80 years in the proportion of tenanted (let) land in the system, a process which has been thought to be accelerating in the last decade and which has, as noted above, given rise to apparent concern. It is difficult to make a clear statement however because distinctions between tenancy and owner-occupation are blurred by modern forms of tenure and management agreements. (For example, it is possible for a father and son partnership to rent from the father, who owns this land, and so appear statistically as occupying a rented holding.) Table 8.1 therefore shows the approximate current position. This appears not to support the contention of accelerated decline. Although the number of rented holdings has halved in the total period shown in Table 8.1, the rate of reduction has not visibly increased in that period, at least in terms of the pattern of holdings. A 'holding' is the basic census unit, but for an examination of ownership and tenure it is rather unsatisfactory since a holding and a farm, as conventionally seen, need not be the same physical or fiscal units. At a higher level still, farms geographically separate can be parts of larger enterprises which are treated as single entities for accounting, investment and taxation purposes. In this sense the Enterprise can be viewed as synonymous with the traditonal Estate, although the physcial form of each will be different. A typical estate might comprise some farm tenancies, an in-hand home farm, with perhaps woodland and sporting rights. Which of these elements might be regarded as subsidiary will depend very much on location. These financially based structures can be very complex, but are of great significance in any consideration of ownership patterns.

Table 8.1: Agricultural Land Tenure in Great Britain, 1950–1984

	Rented/mainly rented		Owned/mainly owned	
	%area	%holdings	%area	%holdings
1950	62	60	38	40
1960	51	46	49	54
1970	45	42	55	58
1980	42	34	58	66
1984	40	31	60	69

Note: Almost all land in Northern Ireland is owner-occupied.
Source: Annual Review of Agriculture, Cmnd Papers.

It is perhaps more important to indicate the structural changes which underlie Table 8.1, and these are shown in Table 8.2, illustrating the extent to which the holding size distribution has changed over the latter part of the same period.

Table 8.2: Great Britain, Number of Holdings by Area, 1973–1983

Hectares	1973[a]	(%)	1978	(%)	1983	(%)
Under 2	20,405	(8.48)	16,371	(7.37)	14,162	(6.53)
2–19.9	88,390	(36.72)	78,974	(35.54)	71,786	(33.09)
20–199.9	123,864	(51.45)	118,152	(53.16)	116,519	(53.71)
200 +	8,078	(3.35)	8,739	(3.93)	14,462	(6.67)
Total	240,737		222,236		216,929	

Note: a. Approximate acreage equivalents.
Source: Agricultural Statistics, United Kingdom, HMSO.

Over the ten-year period shown above, the total number of holdings, as opposed to farms, fell by about 10 per cent. The only size category to experience an increase in holding numbers was that containing the largest holdings, which increased by 79 per cent. By comparison, the smallest size category decreased by 31 per cent, a fact partly attributable to changes in the basis of inclusion over the period. In 1983, the mean size of all holdings in Great Britain was 75.31 hectares. The 6.67 per cent of holdings of more than 200 hectares occupied 47.71 per cent of the total farmed area, with a mean size of 539.1 hectares. By comparison the 6.53 per cent of holdings of less than 2 hectares occupied only 0.09 per cent of the total area, with a mean size of 1.05 hectares. The average area of the largest holdings in the owned and mainly owned group (553.7 hectares) was very similar to the average area of the largest rented or mainly rented holdings (522.05 hectares).

These statistics can only offer a generalised view of the state of land holding because so many changes have had to be made to the bases of the figures in the last 20 years. An example of the difficulty of inter-pretation can be found in the 1983 census of minor holdings. In terms of holding size, the unit of measurement in Table 8.2, there is in any case an overlap between this and the main Ministry of Agriculture (MAFF) census. Holdings of up to 6 hectares may be treated as minor if their overall level of agricultural activity is low. The definition of these holdings has also changed over time, the last change being in 1980, for the countries of the United Kingdom, except Scotland. There were about 43,300 minor holdings in England and Wales in 1983. At the same time however, holdings were moved into and out of this category from the main census of agriculture. The net result was the addition of 4,147 holdings to the main category. The 1983 exercise also removed from the record nearly 15,000 minor holdings which had ceased agricultural activity. With the added complexity of holding/farm occupancy making it almost impossible to produce a set of statistics which enables the struc-ture to be described unambiguously, it is perhaps best to treat the data simply as 'snapshots' of the state of the system at identified points in time.

Why Own Farmland?

In the first instance, land is a factor of agricultural production, and as such is a necessary as well as a 'worthwhile' asset.

Inevitably however this asset is treated as part of the wealth of any owner-occupier or landlord. With an average holding size of about 75 hectares, and an average value close to £3,500 per hectare, the asset value of such a holding is in excess of £250,000. For the owner-occupier that asset value is redeemable by sale, subject to several capital taxes, a Gains Tax on sale, or a Transfer Tax on inheritance. The conventional farmer does not usually view his land-holding in this way, generally not wishing to sell his land but preferring instead to treat it as a factor of production to be kept in good order, like any other capital asset. Land has the further advantage that properly managed it will not depreciate measurably. Land ownership has been traditionally regarded therefore

as a commitment to a way of life and its continuation over many generations. Even periods when the price of land has been relatively high have
not produced higher than average sales. Evidence points to the reverse,
less land coming forward when prices have been high (Jones Lang Wootton, 1983, p. 23).

For a landlord the position is rather different. Agricultural land is
seen as a low risk investment. It is a permanent, non-depreciating feature,
and is in consistent, if not always strong, demand. Because of the low
risk, initial yields have been historically low, from 1–3%, but by comparison captial growth has been high. Jones Lang Wootton report that
between 1961 and 1984 farms have outperformed all other property investment sectors. Overall returns to agricultural property averaged 13.0
per cent in that period (Jones Lang Wootton, 1985).

On traditional let estates there is often a mixing of owner-occupation
with lettings. Even so, landlords of such estates have taken such opportunities as exist to bring previously tenanted land in-hand to farm it
themselves, so reducing the overall number of tenanted holdings. There
has also been a change in the type of tenancy offered on such estates,
with a move from traditional 'model clause' lettings with the tenant
responsible for some repair costs, to full repairing and insuring tenancies where tenants are responsible for all such costs.

Financial institutions have recently found agricultural land offering
attractions as an investment option, although the type and level of interest
will differ from institution to institution depending on investment and
management strategies and abilities. There are several reasons for this
new involvement. The demand for tenanted land has been sustained at
high levels, and land has always been fully let, even if not always with
fully acceptable tenants! Guaranteed and rising product prices have effectively supported an upward movement in rents, and the land itself
in terms of capital growth had kept its values, at least up to the end of
the 1970s.

A well researched study, (Saville/RTP, 1983) has shown that let land
has been an important sector for the institutions generally, and that they
do not appear to be irresponsibly taking land out of tenancy. Indeed there
are distinct advantages to institutions in retaining tenanted farming. First,
there is no need for a specialised management structure devoted
specifically to agricultural property as there has to be for efficient in-
hand farming. Similarly, there is little direct involvement of the kind
required for a partnership approach, and thirdly, any medium or short
term risk as measured by farm income is borne by the tenant.

Savills/RTP show that the traditional estates accounted for 37.9 per cent of the area held by financial institutions in December 1982. This was in spite of the fact that these estates do have a heavy management overhead and may be under-capitalised and in need of rationalisation. They are potentially valuable however, since with proper management they can often be relatively easily upgraded with quite modest new investment to give good returns on capital.

Sale and leaseback has formed the most important way into farming for many institutions. Forty-five per cent of total area bought has been of this type, although recent evidence indicates that sale and leaseback has declined as an institutional option (Farmland Market, 1985).

The third main category has been the purchase of full repairing and insuring tenancies, although this group forms only 15.2 per cent of area.

Institutional interest in farmland dropped away in 1983 and 1984. The difficulties and uncertainties associated with the Common Agricultural Policy (CAP), and the new Agricultural Holdings Act were among the factors responsible. Additionally, the much improved investment opportunities outside property have meant that property in all sectors has become less significant as an institutional investment target. The institutions are now much more selective generally with respect to property, and since in many cases agricultural land has only been included to produce a 'balanced' and complete property portfolio, it is unlikely that they will wish to increase their ownership substantially, but will be prepared to rationalise, if the opportunities present themselves, by judicious sale or purchase.

There is no doubt, however, that from 1974 to the early 1980s, these institutions were the major force in the market, but that was also true for the other commercial property sectors. In agriculture their interest has always been greatest in the largest units and the best quality land and in consequence it seems to be very much the case that the market has come to rely on the insititutions to guide opinion and set price levels, so much so that the present depressed state of the market is at least in part due to the feeling that the institutions no longer wish to participate in the market.

One further ownership group is of some importance in the market for agricultural land, but must also be regarded as having no actual interest in the value of the land for farming purposes. This group consists of the property developers, the most significant of whom in this context are the large residential developers. Urban growth, and housing in particular, has been responsible for the outright loss to agriculture of around 13,000 hectares a year, and the pressure for development land on the urban fringes remains great (Best and Anderson, 1984).

Who Owns Farmland?

The basic answer to this question should by now be apparent, that is, nobody really knows! Northfield identified three major groups of land owners, and indicated the aggregate ownership of each.

Traditional Institutions, Public and Semi-public Bodies

The Church and the Crown are perhaps the two best-known of 'traditional' land-owning institutions, owners who have possessed land over many hundreds of years. Other large owners in this category are the Oxford and Cambridge colleges, other universities and educational establishments. The great majority of their land is let to provide investment income. Central government, through its agencies is a major landowner. The Ministry of Defence, for example, has approximately 110,000 hectares, the Forestry Commission 150,000 hectares. This land is held primarily for operational purposes in the first case, although some is let on a regular basis, and for income in the second case. Nationalised industries, such as the Coal Board and the Regional Water Authorities hold large amounts of land purely for operational reasons, although relatively small amounts are farmed also. Local authorities also own substantial holdings, especially for the creation of tenanted smallholdings, a rather special case, but one which places these organisations in the large public ownership group (CEC, 1981, p. 44).

Northfield estimated, and CAS agreed (Harrison *et al.*, 1977) that the total ownership by this category was of the order of 1.5 million hectares at that time. The figure cannot be very different today, although some public bodies have been encouraged to dispose of surplus land and this has happened in some cases. Table 8.11 indicates that the net loss of land by public authorities since 1977 has been only 15,000 hectares, and that others in this large group have gained 64,000 hectares.

The Financial Institutions

During the 1970s this group of land-owners, consisting mainly of pension funds and insurance companies with some property unit trusts and banks, were the largest net purchaser of agricultural land. It is estimated that by the end of 1982, they held about 400,000 hectares (2.25 per cent) (Steel and Byrne, 1983), much of which seems to have been acquired from individuals who have then leased it back (Savills/RTP 1982).

The proportion of farmland in an institution's portfolio is usually small, and the proportion of institutions having large amounts of farmland is also small, although the correlation between asset size of institution and asset value of farmland is low (Steel and Byrne, p. 15). At least 80 per cent of land held in 1982 was let by one means or another, the rest being in vacant possession, farmed in-hand, mostly by the largest institutions.

Private Individuals, Companies and Trusts

The remaining land, by subtraction from the total, 15.8 million hectares, is in the ownership of private individuals, companies or trusts. Very little is known of the detailed structure of ownership as between these interests. A sample survey conducted in 1976 by the Agriculture EDC (AEDC, 1977) covered about 18 per cent of the total when it looked at 1,677 estates in England and Wales, and found that 38.1 per cent by area was in individual or joint ownership, 33.6 per cent was held by trusts of various kinds, 2.3 per cent by family companies, and 19.1 per cent by the ubiquitous institutions.

The trust form of ownership is in many cases a proxy for personal ownership, a device intended to assist in inheritance transfer and lifetime tax-minimisation. This being so, in 1976 nearly three-quarters of tenanted land was effectively in personal ownership.

Agricultural Land Price Statistical Series

Although systematic data on tenure are not available the database for prices of land is rather better, but variety is again an overriding theme.

Agricultural land is sold in one of two ways, by public auction or by private treaty. The majority of statistical series relate to publicly available auction data, since the results of private sales tend to remain private for the most part (but see Munton, 1975). It must also be borne in mind that in any one year only a very small proportion of the total of land is offered for sale. In England and Wales in 1983 for example, about 152,400 hectares was sold with vacant possession, and 16,600 hectares of tenanted land, about 1.75 per cent of the total land area (ADAS series).

There is only one series based on the results of all sales, including private treaty sales, and this is compiled from the stamp duty returns of the Inland Revenue, and published by the Ministry of Agriculture, Fisheries and Food (MAFF). All transfers of land of five hectares or more are included. As an up-to-the-minute indicator of activity in the market, however, the series has a major shortcoming, especially in periods when the market is in any state of flux. Based as it is on returns made for duty purposes, it uses as its time base the point at which the duty is paid rather than the point at which the sale is agreed. This would

Table 8.3: Crude Yearly Average Land Prices in England and Wales, 1975–1984 (Year ending in September)

All sales of more than 5 hectares

Year	Price/hectare (£s)
1975	1,214
1976	1,079
1977	1,291
1978	1,802
1979	2,316
1980	3,039
1981	3,162
1982	3,098
1983	3,321
1984	3,631

Note: Although these figures are related to the year of publication in this table, they are usually interpreted as having a nine-months' lag, and so refer to the previous calendar year.
Source: MAFF/Inland Revenue.

obviously not be a problem if all sales proceeded at the same rate, but the lag is very variable both within and between transactions, averaging about nine months. The series is therefore limited to providing an historic overview based on all data. As such, it has some value in showing the underlying trends in the market. Land prices from the series for the decade up to 1984 are given in Table 8.3.

While these figures are not adjusted for inflation an upward trend in price is apparent, as is a downturn in the mid-1970s following on the collapse of all sectors of the property market in 1974 (Farmland Market, 1975), and uncertainties of the market in the last few years (Farmland Market, 1985).

A more up to date series has been published since 1976, based on sales data collected by the Agricultural Development and Advisory Service (ADAS) of MAFF, and by the Agricultural Mortgage Corporation (AMC) which provides mortgage services for the purchase of farmland. In 1980 they were joined by the Country Landowner's Association (CLA), who had produced a series based on returns made by chartered surveyors and land agents in England and Wales. This series attempts to report both vacant possession and tenanted sales in England and Wales, and covers about 40 per cent of all land sold. Usually transactions are included in this series within three months of occurence. The level of prices recorded by this survey is shown in Table 8.4, which also makes clear the marked difference in price level between vacant possession and tenanted land which was not apparent in the Inland Revenue-derived prices of Table 8.3.

Table 8.4: Current Agricultural Land Prices 1975–1984:
Quarterly Rolling Average Price per Hectare (£s)[b]

All sales of more than 5 hectares.

| | Vacant Possession | | Tenanted | |
Year	England	Wales	England	Wales[a]
1975	1,281	—		
1976	1,614	—	904	—
1977	2,056	1,554	1,135	—
1978	3,249	2,168	1,406	—
1979	4,156	3,007	2,868	—
1980	3,980	3,208	2,558	—
1981	3,988	2,732	2,535	—
1982	4,705	2,929	2,009	—
1983	4,700	3,210	2,378	—
1984	4,202	3,604	2,399	—

Notes: a. Welsh tenanted land averaged 14 sales per annum in the period
1977–1983, at about £1,050 per hectare. b. Figures relate to the third
quarter of each year.
Source: ADAS/AMC/CLA land price series.

In as much as information is available on sales of tenanted land, the
vacant possession premium for sales in this series during 1983 averaged
just under 50 per cent. This is however based on a very small sample,
and indeed the Oxford Institute series (see below) reports insufficient
sales for the five years 1980–85, to allow averages to be meaningfully
calculated.

The third main source of land price data for England and Wales, is
the long running series which originated in the Institute of Agricultural
Economics at Oxford and which has been continued in collaboration with
two professional journals (Estates Gazette and Farmers Weekly) as 'The
Farmland Market'. The data derives from published reports of auction
sales, with extensive coverage by both journals to ensure as complete
enumeration as possible. Inevitably some sales are missed, but their
significance to the market as a whole is probably very small. Because
auction sales are only part of the market, there may be a resultant bias.
This is difficult to assess, and it is clear that these data are widely used
as the basis for commentaries on market performance. The raw data are
published, and Table 8.5 indicates the overall performance of the series
in the last ten years.

When the effects of inflation are taken into account the average performance of the market takes on a rather different picture. After reaching
a peak in 1979, the real price of land has recently fallen to the same general
level as in 1977. The overall trend in prices remains upward, but the
variability of the series from year to year is quite large. (Table 8.6).

Table 8.5: Auction Prices in England and Wales 1975–1984:
Median Price per Hectare

Year	Farms with possession Price (£)	No. of Sales	Bare land Price (£)	No. of Sales
1975	1,484	334	1,484	662
1976	1,964	467	1,905	799
1977	2,528	532	2,412	1139
1978	3,383	466	3,353	1193
1979	4,517	437	4,495	1251
1980	4,705	283	4,487	832
1981	4,501	307	4,491	986
1982	4,813	363	4,745	1103
1983	5,452	371	5,190	1272
1984	5,481	267	4,975	947

Notes: Farms approximately 10 hectares and more. Bare land approximately
2 hectares and more.
Source: The Farmland Market.

Table 8.6: Mean Land Prices Adjusted For Inflation[a]
1975–1984, England and Wales: Farms and Bare Land,
Vacant Possession Auction Price per Hectare

Year	Price (£)
1975	1,332
1976	1,555
1977	1,813
1978	2,243
1979	2,636
1980	2,180
1981	1,952
1982	1,917
1983	2,076
1984	1,873

Note: a. Retail Price Index used as Inflation Measure (1975 = 100).
Source: Oxford Institute Series.

Types of Agricultural Land

Before proceeding to a discussion of structure and patterns in the
agricultural land market, a brief description should be given of the basic
commodity under discussion — agricultural land. Although there are other
variable factors which influence the value of a parcel of land, a major
factor will always be the land quality. There are five recognised categories
of agricultural land quality in England and Wales, and a different, if

related, system in Scotland. Land is categorised particularly by soil type but also according to altitude, climate and relief. These factors are then related to the extent by which they limit agricultural usage, to arrive at an overall grade. The gradings are as follows:

Grade 1. Best quality land. Cropping is highly flexible with few restrictions as to crop type and few, if any, physical limitations.

Grade 2. Land with a few limitations, but generally very fertile. Most crops can be grown, but there are limits on horticulture.

Grade 3. The average category, covering a wide variety of combinations of soil, relief and climate. There are restrictions on use, especially on horticulture and root crops. Grass and cereals are usually the main crops.

Grade 4. Lowest quality capable of reasonable cultivation. There are often quite serious limitations due to physical factors, and output is confined to pasture or rough grazing.

Grade 5. This land is marginal to agriculture, having very severe limitations. This land is almost always under rough grazing, or alternatively forestry planted land.

Within the UK there are wide differences in the proportions of land in each of these grades as is shown in Table 8.7.

Table 8.7: Agricultural Land Quality in the United Kingdom, Thousands of Hectares

Grade	England	Wales	Scotland	GB	NI	UK
1	326	3	20	349	1	350
2	1,653	36	155	1,844	38	1,882
3	5,345	295	880	6,520	455	6,975
4	1,554	746	660	2,950	529	3,489
5	1,019	604	4,765	6,388	62	6,450

Source: MAFF 1978.

On the ground, grades of land are very mixed. Generally, though, the greatest proportion of better quality land is to be found in the Eastern counties of England and Lowland Britain (Best, 1981, p. 145).

There are very clear differences in the attractiveness to purchasers of these varying land types. This has a marked effect on the pattern of land prices and values, as is shown in Table 8.8.

Table 8.8: Average Price of Agricultural Land by Land Class
for England and Wales 1983/84 (£/hectare)

Grade	England	Wales	% Area
1 & 2	4942	4801	13.96
3	3859	3816	54.70
4 & 5	2108	2416	31/34

Souce: Agricultural Land Prices in England and Wales.

The majority of sales are for land of mixed type, and the prices
achieved are related to the predominant land quality. Some sales comprise land and buildings (about 36 per cent in England and Wales in 1983),
and in some parts of the country, especially the South and South-east
of England, the inclusion of a good quality farmhouse and buildings and
the proximity of non-farming amenities can add substantially to the value
for some purchasers.

The Pattern of the Agricultural Land Market

Although the data to permit a detailed description of current ownership
are effectively non-existent, ADAS have since 1978 produced an il-
luminating series based on Inland Revenue returns showing the marginal
changes in ownership for six main categories. The categories of ownership overlap to some extent, a function of the basic complexity of methods
of ownership and of the fact that placement in a particular group is according to the opinion of the District Valuer who makes the return, but
they provide the clearest picture of the dynamic of the market since their
introduction.

Table 8.9 shows the form which these statistics take, as at 30
September 1984, but with the lag back to calendar year 1983.

These groupings — property companies, financial institutions and
other companies — are amorphous to the extent that identification of
the type of vendor or purchaser can sometimes prove difficult and they
may therefore by misassigned. Generally however the District Valuer
is able to asssign a transaction correctly and the misplacement is marginal.

By far the largest numbers (and area) of transactions are between individual farmers in the vacant possession market, about 70 per cent of
the total area traded. The other categories of purchaser operate therefore
on the margins of this market, but their effect has been disproportionately
large, especially in terms of the prices made for some types of farm in

Table 8.9: Sales of Agricultural Land by Type of Vendor and Purchaser, All Properties, 1983/84, England (Wales) 5 Hectares and more

	No. of transactions		Aggregate area, hectares		Av. price £/hectare	
Type of vendor						
Individual	3,942	(922)	103,755	(21,430)	3,831	(2,908)
Property company	76	(18)	2,557	(305)	3,112	(2,238)
Financial institution	94	(20)	9,101	(1,070)	3,520	(2,305)
Other company	353	(10)	18,890	(455)	3,123	(2,432)
Public authority	134	(40)	5,351	(1,498)	2,403	(724)
Other	90	(10)	4,271	(347)	3,097	(2,351)
Type of purchaser						
Individual	4,093	(996)	103,547	(23,346)	3,681	(2,741)
Property company	71	(5)	4,499	(139)	4,597	(5,124)
Financial institution	39	(a)	9,205	(a)	3,474	(a)
Other company	402	(16)	20,781	(1,228)	3,533	(2,495)
Public authority	72	(6)	2,158	(185)	3,579	(994)
Other	66	(5)	3,735	(213)	2,017	(2,526)

Note: a. No sales, or insufficient sales to avoid confidentiality constraints.
Source: ADAS/Inland Revenue.

some areas.

The transactions are also broken down by tenure and area, and indicate that in England (1983/84), 95.07 per cent of transactions, occupying 88.55 per cent of the area were for fully tenanted land. Not only is there a difference in price between vacant possession and tenanted land, but the differential is extended according to size of holding, with, as might be expected, larger units being cheaper per hectare. (Table 8.10).

Table 8.10: Average Price and Number of Sales by Size of Transaction, 1983/84, England (£/Hectare)

Size Group	Vacant possession		Tenanted	
	No.	Av. price	No.	Av. price
5– 9.9	1,778	4,656	46	2,733
10–19.9	1,129	4,461	36	2,809
20–49.9	987	4,151	57	2,305
50–99.9	391	3,780	48	2,179
100 +	173	2,990	44	2,462

Source: Agricultural Land Price Statistics.

The simplest, indeed the only way of discovering the effect of any group within the market as a whole over time is by examining the net sales or purchases for each ownership group. Each group has transactions both as seller and buyer. Data are available in aggregate form from 1977, and the net effects are shown in Table 8.11.

Clearly land has consistently moved out of individual ownership, and into some form of corporate ownership. It would seen most unlikely that the owners of 111,900 hectares of land have left farming altogether, and the conclusion must be that the majority of these sales have been on the basis of an agreed leaseback to the owner by an institution. At the end of 1982, 64 per cent of institutional purchases had become leasebacks, and just over 83,000 hectares were under such tenancies (Savills/RTP 1982). The yearly purchases of the majority of that area are therefore to be seen in the table. It must be remembered that in every year the institutions also sold land, and there were structural changes taking place in the distribution of their holdings. Part of this can be seen in their effects on the tenanted market in England.

Here the majority of purchases are always by sitting tenants. In 1983, for example, of 231 tenanted transactions, 184 were acquired by

Table 8.11: Net Land Acquisition by Type of Owner, 1977–1983, England (Thousands of Hectares)

Type of Owner	1977	1978	1979	1980	1981	1982	1983	Net
Individual Property	−23.7	−15.2	−35.2	−15.7	−13.9	−8.0	−0.2	−111.9
Company	3.7	1.3	2.0	1.4	2.6	−0.3	1.9	12.6
Financial Institution	14.1	9.7	12.0	10.9	9.1	2.5	0.1	58.4
Other Company	8.4	9.6	9.9	4.6	4.2	4.2	1.9	42.8
Public Authority	−1.0	−1.1	1.3	−0.9	−4.4	1.4	−3.2	−7.9
Other	−1.4	−4.3	10.1	−0.9	2.3	0.1	−0.5	5.4

Note: Figures are rounded and therefore do not sum exactly.
Source: Agricultural Land Prices in England and Wales.

individuals, with a mean transaction size of 55.6 hectares. By comparison, institutions and property companies had a gross acquisition of 3,381 hectares of tenanted or leaseback land in only 13 transactions (mean size 260.1 hectares). Their net acqusition of tenanted land in 1983 was however only 302 hectares and this further implies substantial local changes in structure.

Additional information on the tenanted market comes from a survey of land loss in the sector made annually by the Central Association of Agricultural Valuers (CAAV). Taken together the results for 1983 and 1984 cover 94,872 hectares. Of this, 57,508 hectares which had previously been tenanted and in private ownership became available in the two years. Less than half, 24,477 hectares (42.56 per cent) was reported as relet. The remaining 33,031 hectares was taken in-hand by one means or another, either to farm or for eventual sale. A further 12,036 hectares was sold to sitting tenants, and was thus also lost to the sector.

The institutional effects noted by the survey are summarised in Table 8.12.

Table 8.12: Institutional Let Land, 1983–1984

Hectares

Financial Institutions	
Fresh lettings of previously let land	4,218
Land previously let, now taken in-hand and farmed in-hand by joint ventures.	2,134
Traditional Institutions	
Fresh lettings of previously let land	4,736
Fresh lettings of land previously farmed by the owner in-hand, or bought with Vacant Possession immediately before creating a tenancy	4,409
Joint ventures on land previously farmed in-hand	9,631

Source: CAAV Survey.

Still further evidence of these structural effects is to be found in the changes in the mean size of transaction for both buyer and seller (Table 8.13). For this analysis, institutions and property companies form one group. In any one year the figures will be governed to a certain extent by the way in which holdings of different sizes reach the market, and the way in which the holdings are disposed of, in lots or as single units. With this proviso, it is interesting to see the consistent average size of purchases by individuals. This is not totally unexpected in the sense that interest in these relatively small units is likely to be largely confined to this group, except in those cases where insititutional or other holdings would be consolidated as the result of land bought. Another notable feature is the steady reduction in the average size of individual sales and purchases.

Most striking is the annual disparity between the selling and buying of the institutional sector. For most years the mean size of purchases has been approximately double that of disposals. This clearly reflects the desire of these organisations, when possible, to acquire effective holdings of sufficient substance to make modern farming methods and economies of scale a practical aim. The number of purchases has fallen steadily from more than 280 in 1977 to 110 in 1983. By implication the institutions have become much more selective in the properties that they choose to buy. Although the number of properties sold has remained much more constant, between 170 and 200, the average size of disposals has been higher on the whole than for either of the other ownership groups.

Table 8.13: Mean Size of Transactions, Sellers and Buyers, 1977–1983, England (Hectares)

	Individual	Institutions	Other
1977			
Selling	30.12	40.79	48.08
Buying	25.46	106.00	49.19
1978			
Selling	31.26	56.33	42.11
Buying	27.19	103.79	49.39
1979			
Selling	31.57	37.17	43.20
Buying	24.23	89.94	67.49
1980			
Selling	27.02	37.84	41.59
Buying	23.39	109.60	40.31
1981			
Selling	27.56	55.17	43.32
Buying	24.43	103.05	46.74
1982			
Selling	26.56	62.08	34.13
Buying	24.52	88.63	45.27
1983			
Selling	26.32	68.58	49.41
Buying	25.64	124.58	49.40

Source: Agricultural Land Prices in England and Wales.

Table 8.14: Net Gains/Losses of Agricultural Land by Ownership Groups and Region 1977–1983, England (Hectares)

MAFF Standard Region/ Standard Statistical Region[a]	1977	1978	1979	1980	1981	1982	1983
INDIVIDUALS							
Northern	-2,491	-281	-14,665	-1,084	-1,028	82	1,442
Yorkshire	-390	317	-3,429	-1,736	-3,050	-3,196	-534
East Midlands	-3,917	-3,492	-2,258	-3,529	-3,553	-894	-601
East Anglia	-882	-138	-1,186	-6,356	-1,331	-3,293	-1,986
South East	-9,148	-7,205	-6,985	-861	-3,458	-711	-197
South West	-3,859	-4,209	-3,608	-1,517	-1,241	-1,396	1,554
West Midlands	-3,056	-212	-2,726	-232	-861	582	108
North West	—	—	—	-149	444	972	9
INSTITUTIONS							
Northern	1,580	550	746	{ 1,222	-1,216	50	{ 1,142
Yorkshire	-527	1,182	1,658	{	811	1,593	{
East Midlands	5,164	4,359	2,263	4,400	5,473	1,470	-855
East Anglia	390	179	430	3,985	1,140	3,779	1,996
South East	6,598	3,279	5,992	1,585	3,578	989	152
South West	2,940	1,789	2,122	821	381	1,918	-931
West Midlands	1,650	-321	756	{ 905	{ 1,554	{ 2,335	140
North West	—	—	—	{	{	{	404
OTHERS							
Northern	913	269	14,278	{ 1,579	2,245	25	-2,256
Yorkshire	917	-1,499	1,771	{	2,237	1,602	516
East Midlands	-1,249	-866	-5	-869	-2,120	-576	1,457
East Anglia	493	-40	755	2,381	189	2,638	-8
South East	2,550	3,925	993	-274	119	880	45
South West	918	2,421	1,486	697	859	1,575	622
West Midlands	1,406	534	1,970	{ 721	{ -1,142	{ -403	-248
North West	—	—	—	{	{	{	-413

Note: a. Regional base of data analysis was changed after 1979 from MAFF Standard Regions to Standard Statistical Regions. There is some overlap.

In almost all years and regions, the institutions are obvious net purchasers of land (Table 8.14). Also observable on a regional basis is the rapid slowdown in institutional buying shown nationally in Table 8.11 which has so disturbed the market recently. As indicated earlier, there are good reasons why this has been so, but the institutions have remained interested in areas of good quality land. There is little doubt that sale and leaseback under the 1976 tenure rules made this an attractive option for many owners, particularly at the prices which the institutions were prepared to offer in the late 1970s. It is inevitable therefore that great weight is given to the institutional and corporate effects on the market during that period. The statistical material does not indicate any specific responses to institutional interest by particular sellers, but anecdotally there was certainly an expectation, largely sustained, that for quality land and large units the institutions were the buyers of first resort.

Conclusion

The last ten years or so have seen strong political support for the agriculture industry in Britain associated with the full implementation of the Common Agricultural Policy. The EEC, through the CAP and other Directives, has provided the incentive for amalgamation of holdings in both the owned and tenanted sectors with its emphasis on the need for flexibility and production. Price support has encouraged farmers to seek the largest possible holdings, in order to maximise net income. This in turn has pushed the price of land of all types and qualities higher than might otherwise have been expected and land of the highest quality disproportionately higher still. Possibilities for the future are complex and uncertain. They depend to a great extent on the performance of the industry being sustained in an economic environment which has been heavily supportive, perhaps over-generous, but which may not remain so. The technological changes in agriculture in Britain in the last ten years have been considerable; there is no reason to suppose that changes in the next ten years may not be equally extensive.

The debate about ownership and tenanted farming will remain unresolved until adequate data are available to provide a proper basis for discussion. What is not in question is the continuing steady loss of land from the tenanted sector. Sales to sitting tenants obviously contribute in part to this loss. Such information as there is at present points to the institutions preferring tenanted farming, quite evidently seeking maximum (but low) returns for minimum additional investment. Given the yields on agricultural property, and the constraints of the tenure system until recently, little else is to be realistically expected. Private owners of tenanted holdings, equally constrained by the legal position, seem much more disillusioned with tenanted land, preferring instead the much greater freedom of use and disposal and the higher capital value of vacant possession land. The CAAV survey has consistently shown that the majority of the loss, however imperfectly measured it may be, is attributable to the private landlord.

The role of the institutions in the agricultural land market is likely to remain significant, at least in the tenanted sector. This is because the institutions will continue to buy and retain prime agricultural property and will prefer to lease it. Because after purchase agricultural land can become an almost costless investment, they have no reason to dispose of very much of their agricultural portfolios as long as the relatively illiquid funds are not positively needed for other investments. In time therefore the highest quality land and perhaps the largest holdings will be in the tenanted sector, although with a much smaller number of mainly institutional owners. This does not improve the prospects for new entrants to farming and certainly does not point to any increase in the proportion of tenanted holdings, rather to a stable point being reached, as Northfield suggested (Northfield, 1979, paras. 110–113), somewhere in the early twenty-first century. If the tenanted market is to be sustained with a reasonable proportion of private landlords, stronger incentives will have to be provided for those landlords to maintain existing tenancies and to offer new land for letting or the sector could well share the fate of the now almost non-existent residential rented market.

The apparent state of land ownership is the cause of so much debate in Britain, basically because we remain in almost complete ignorance of the fundamental structures of ownership, so that the real consequences of new developments taking place in the system are difficult to measure and interpret in any but the broadest of ways.

References

ADAS (1973) *Agricultural Land Prices in England and Wales and the Construction of Land Price Indices,* Technical report 20/6, MAFF, Pinner, Middlesex

ADAS (Annual) *Agricultural Land Prices in England And Wales,* MAFF, Alnwick

Agriculture Economic Development Committee (1977) *The Ownership of Land by Agricultural Landlords in England and Wales,* National Economic Development Office, London

Best, R.H. (1981) *Land Use and Living Space,* Methuen, London

218   *Land Ownership and the Agricultural Land Market*

Best, R.H. and Anderson, M. (1984) 'Land-use structure and change in Britain, 1971 to 1981', *The Planner, 70* (11), 21–4

Collins, T. and Smith, L. (1979) *Land Parcels in a Computerised Land Information System*, Proceedings of the 8th Australian Computer Conference, Canberra

Commission of European Communities (CEC) (1981) *Factors Influencing the Ownership, Tenancy, Mobility and Use of Farmland in the United Kingdom*, Information on Agriculture No. 74, EEC, Luxembourg

Estates Gazette/Farmers Weekly (February/August) *The Farmland Market*. Estates Gazette, London, and Farmers Weekly, Sutton, Surrey

Farmland Market (1975) 'Why have farmland values fallen?' *The Farmland Market, 3,* Jan., 9

Farmland Market (1985) 'Land values 1984', *The Farmland Market, 23,* Feb. 4–5

Hansard (1983) *Debate on the Agricultural Holdings Bill (Lords)*, 13 Dec. 1983, cols. 206–7, HMSO, London

Hansard (1984) *Debate on the Agricultural Holdings Bill (Lords)*, 6 June 1984, cols. 333–45, HMSO, London

Harrison, A. Tranter, R.B. and Gibbs, R.S. (1977) *Land Ownership by Public and Semi-public Insititutions in the United Kingdom*. Paper No. 3. Centre For Agricultural Strategy, University of Reading, Reading

Inland Revenue (Bi-annual) *Valuation Office Property Market Report*. Surveyors Publications, London

Jones Lang Wootton (1983) *The Agricultural Land Market in Britain*, Jones Lang Wootton, London

Jones Lang Wootton (1985) *Returns to Property: Analysis of the PPAS Database*, Summer 1985, Jones Lang Wootton, London

Kalms, T. (1985) *Land Registration within the Land Information System*, Paper given at the joint Urban Data Management-Spatially Oriented Referencing Systems Association Symposium, June 1985, The Hague

Ministry of Agriculture, Fisheries and Food (MAFF). (Annual) *Agricultural Statistics, United Kingdom*, HMSO, London

Munton, R.C.J. (1975) 'The state of the agricultural land market 1971–73', *Oxford Agrarian Studies. 4,* (2), 111–30

Northfield (1979) *Report of the Committee of Inquiry into the Acquisition and Occupancy of Agricultural land*, Command 7599, HMSO, London

Royal Institution of Chartered Surveyors (1977) *The Future Pattern of Land ownership and Occupation*, A discussion paper of the Land Agency and Agriculture Division, Royal Insititution of Chartered Surveyors, London

Steel, A. and Byrne, P.J. (1983) *Financial Institutions. Their Investments and Agricultural Landownership*, Working Papers in Land Management and Development, No. 1, Department of Land Management and Development, University of Reading, Reading

Savills/Roger Tym and Partners (1982) *The Savills-RTP Agricultural Performance Analysis.* Savills/RTP, London

Note: In addition to the citations above, leading firms of Land Agents produce various reports on the state of the market at regular intervals, Humbert's Commentary and Savill's Agricultural Land Market Report being but two examples. The Estates Gazette has a regular monthly column, Country Practitioner, with useful data, and reviews the sector annually in January. The Chartered Surveyor Weekly also includes current items of interest on the state of agriculture and the markets for land, as do Farmers Weekly and Country Life.

9 AGRICULTURAL MARKETING AND DISTRIBUTION

W. Smith

Over the past 50 to 100 years, the links binding farmers to consumers have been subject to a variety of pressures that have radically restructured agriculture. Farmers, threatened with an increasingly passive role in the food system, have demanded government intervention on their behalf. For the most part, governments have complied, thereby further restructuring the agricultural market system. Structural changes and the regrouping of retailing, processing, and the farm-supply industries directly affect farmers. For agricultural geographers, the nature and impact of these changes are among the most significant emerging research issues (Bowler, 1984).

The links between farmers and their markets shape the economic organisation of the rural landscape. These links not only permit the exchange of goods and services, but also affect the structure and scale of farm operations and the rate and direction of technological change (Smith, 1974). Changes in market structure can fundamentally alter the organisation of agrarian society (Skinner, 1967). Consequently, the importance of agricultural marketing goes beyond the mere logistics of transportation — it impinges on basic sources of political conflict, social justice, and economic power (Brown, 1984).

This chapter examines the changing characteristics of agricultural markets and explores the dual role of the market as a means to transport, process, store, and grade produce, and as an information system that connects produce with consumers in terms of place, time and form. These roles make the market system a key instrument for imposing change on the farm community, change generated for the most part by urban needs.

Agricultural geographers have contributed little to current market issues, particularly the problems facing agriculture in the industrialised world. Efforts to apply existing spatial theories about markets reveal incomplete analyses of (1) the complexities of demand and (2) the nature of national agricultural policies. Removal of these shortcomings would strengthen the contribution of geographers to rural planning, policy, and development issues.

Models of Agricultural Markets and Land Use

Analyses of the relationships between farmers and their markets can be traced back to the work of Johann Heinrich von Thünen in 1826 (von Thünen, 1966). The theoretical ideas he expounded rest on the essential relationship between transportation costs and land use (Figure 9.1). Using this thesis, he described an imaginary landscape centred on a large town, on a plain of uniform fertility, in which production costs were proportional to distance. Von Thünen reasoned that production costs decrease with distance whereas transportation costs increase. In consequence, items that are heavy or bulky in relation to their value should be produced near the town since it would be more expensive to transport them from more distant areas. Perishable goods should be grown near the town to avoid losses from spoilage. Land more distant from the town should be used for providing goods that are lighter in relation to their value and can therefore justify higher unit transport costs.

Figure 9.1(a): Theoretical Situation for Three Products assuming Constant but Different Freight Rates

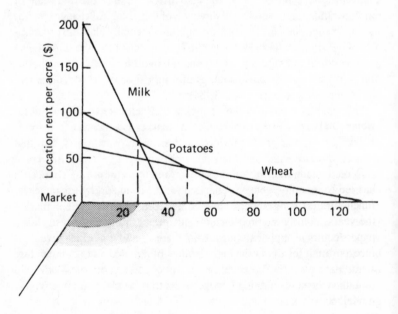

Figure 9.1(b): Von Thünen's Concentric Agricultural Zones around an Isolated Town on a Uniform Plain

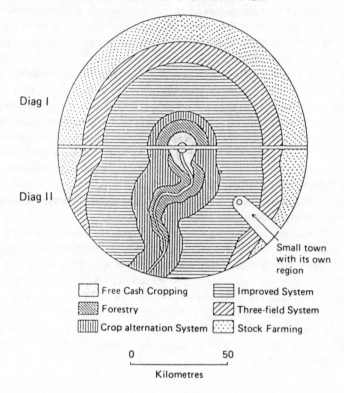

Notes: Diagram 1: General Case
Diagram 2: With navigable river and a small town with its own region.

Von Thünen laid the foundations for the succession of studies that has led to a comprehensive understanding of the spatial structure of a functioning economic system (central place theory) (Johnson, 1970). Over the years, many geographers have elaborated on von Thünen's work; papers continue to appear investigating detailed aspects of his theory and incorporating missing variables into his model of the agricultural landscape (Cromley, 1980; Huff, 1981; Jones, 1982; Visser, 1980).

Little can be accomplished from the repeated testing of the von Thünen model. Its application to different physical environments, cultures, and scales of analysis is well known (see, for example, Chisholm, 1962). The continued emphasis on this static approach has discouraged dynamic analyses that involve both spatial patterns and the factors responsible for their evolution (Norton, 1979; Smith, 1984). However, no satisfactory alternative framework has emerged that incorporates the complex characteristics of the food market in the modern world — a market in which farmers, processors, distributors, retailers and consumers operate as parts of a single system.

Geographers and other social scientists who recognize the importance of the market in determining land use rely on models that examine 'horizontal' competition within a given market (Found, 1971; Gregor, 1970; Morgan and Munton, 1971), for example, the competition among different retail companies for consumer business in one urban region. This emphasis on competition at one particular level in the market system ignores the fact that competition also takes place at other levels (and between levels), that within any one level competition is limited to a specific range of goods, and that in a modern economy there are wide areas of monopoly control. Often competition is among only a few major actors.

Belief in a simple balance of supply and demand, based on assumptions of price, comparative advantage and perfect competition is no longer valid. Equilibrium models imply equilibrium pricing. They do not take into account 'vertical' relationships — those linkages between different levels in the market system — for example, the relationship between farmers, processors, and retailers. Normative models of rural land use examine one level in the market after making assumptions about other levels. As a result, they are much less comprehensive than is often implied (Smith, 1984).

Market Structure

All market systems share similar characteristics, encompassing the movement of goods from the farmer to processors and wholesalers, or directly to individual consumers. In Western economies market links 'upstream' and 'downstream' from the farm sector have been growing in strength since the mid-1940s. A similar development is now occurring in Eastern Europe and the Soviet Union (Hebden, 1984), and in the Third World (Wallace and Smith, in press). It is, however, important to be cautious in extrapolating evidence from the industrialised world to Third World contexts. Even within the Third World, marketing systems exhibit a range of distinctive characteristics related to specific national factors (Kaynak, 1981).

Whether farm produce moves directly from the farmer to the consumer

Figure 9.2: Stages between Production and Consumption

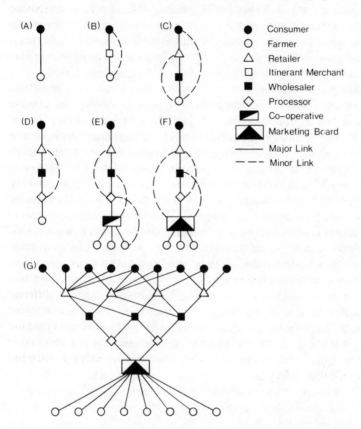

(Figure 9.2 (a)) or through a large number of intermediaries (Figure 9.2 (f)), the process is similar — only the number of market agents and their degree of specialisation change. Farmers who sell produce direct to the consumer fulfil the joint roles of producer, processor, wholesaler, and retailer. This occurs, for example, when dairy farmers process their own milk into cheese, subdivide it into amounts suitable for individual households, and sell it to the households themselves.

As an economy grows, the market chain tends to lengthen and the market agents become increasingly specialised. However, the transition from the model shown in Figure 9.2(a) to that in Figure 9.2(g) is not inevitable. In fact, in recent years in some developed countries, direct contact between farmers and consumers has increased (Bowler, 1981; Shakow, 1981). Even the role of the itinerant merchant in the Third

World (Smith, 1979, 1980) persists in industrialised countries in the modern salesperson and buyer. However, most food in the industrialised world, and an increasing proportion in the Third World, is bought by consumers from retailers, many of whom also act as wholesalers. In addition, a processing sector is now part of many market systems. Throughout the world, the penetration of agriculture by corporate capitalism is increasing.

For any specific item, the market system may offer farmers many possible outlets. Therefore farmers producing more than one item may participate in a large number of market relationships. However, the market options available to farmers are decreasing (Jumper, 1974). Fewer market options means less price competition. Although marketing boards (Figure 9.2(c)) and co-operatives (Figure 9.2 (f)) may take on the roles of wholesalers and processors and so increase farmers' bargaining power, at the same time supply management regulations may decrease a farmer's market options by requiring that farm produce be sold through a statutory marketing board (Figure 9.2 (g)). The only producers exempted from these regulations are those too small to be covered by government controls or those who sell illegally.

A farmer's choice of market may also be restricted if any one corporation offers the sole market outlet at any level in the market system. The geographical monopoly held by a large co-operative or processing firm in a particular area is a case in point. Consequently, the market structure for many items may be envisaged as two pyramids one inverted over the other, with power concentrated at some point between farmers and consumers. In such a situation, farmers risk losing their individual freedom and are vulnerable to the misuse of corporate power (Breimyer, 1965).

Market Integration

The dominant trend throughout the market system in recent years has been the increasing concentration of functions through both vertical and horizontal integration. For example, some processing firms have carried out vertical integration by extending control backwards into farming and forwards into wholesaling, retailing, and even into the restaurant business. Food processors, wholesalers, and retailers have also extended their control horizontally, not only to obtain a larger share of a specific product market, but to gain a share of the market for other food products. For example, canned food manufacturers have expanded into bakery goods and frozen foods.

Whereas the lack of market integration in the Third World has prompted concern as a barrier to agricultural development (Luqmani and Quraeshi, 1984; Mittendorf and Abbott, 1979), market integration is well advanced in western, industrialised countries, where it was encouraged by a radical reorganisation of the retail industry initiated in the early 1950s (see, for example, Heflebower, 1957). This came about as a direct, competitive bid by retailers to profit from the perceived latent demand for lower-cost goods sold from centralised sites with few services attached. Concentration on quick, bulk cash sales without delivery and with less prompt service has allowed mass distributors to survive on lower profit margins at the expense of smaller traditional retail stores. This has provoked structural changes throughout the market system and redistributed market power.

The full transformation of the agricultural market is only possible in societies in which consumers are mobile and can store food in bulk at home. Consequently, the integration of food retailing and wholesaling started first, and has proceeded furthest, in North America, where most families have a refrigerator and own a car. Improved communications and transportation techniques, blatant use of mass advertising, and consumers' willingness and ability to pay for processed food have combined with the demands of the reorganised retail sector for regular supplies of products of uniform quality and type to increase concentration throughout the distribution system. In Canada, for example, the top four companies in some key sectors of the food processing industry share over 90 per cent of the market. National data downplay the problem, since the real consumer market is the urban area. A review of 32 Canadian urban centres in the 1970s found that the market share of the four largest retail companies in each centre was as high as 98 per cent (Mallen, 1976). Despite substantial changes in corporate ownership, the overall levels of concentration have probably changed little since then, if they have not increased.

Corporate conglomerates are getting increasingly involved in every area related to agriculture both in industrial countries and in the Third World (Cracknell, 1980). A few companies dominate more than one sector of the food industry. These conglomerates have extended their control up and down the market chain that links farmers to consumers and well beyond the limits that their original product base might suggest. For example, some North American food retailers have expanded into sugar refining and flour milling. Other companies have chosen to extend control laterally as, for example, when food retailers diversify by taking over furniture stores, or when bakery goods manufacturers expand into frozen vegetables or canned meat. This type of expansion has occurred at all levels in the market system from retailing through to wholesaling and processing. These changes give individual corporations vast decision-making power, while they minimise risk by diversifying a corporation's products and increase the flexibility of its financial planning and control.

The concentration of decision-making power at every level in the market system has endowed agriculture with many industrial characteristics. For many farmers, the freedom to make their own decisions is limited by corporate control and market pressures. However, despite the divisions within farming created by varying income levels and scales of production (Ehrensaft, 1983; USDA, 1981; Wallace and Smith, in press), agriculture remains for the most part a sector of numerous, scattered, small firms. This situation, made even more complex by agriculture's dependence on the physical environment, makes it difficult for the market system to respond quickly and effectively to the diverse demands of modern consumers. It also compounds the farmer's vulnerability in the market chain.

Hedging Market Risk

The production and supply of food for the market involves a high degree of economic risk and uncertainty. A range of risk avoidance strategies has evolved at every level in the agriculture-food system to counter these threats. The attempt by corporations to improve profits and stability by controlling a large market segment underlies the concentration of the retail, wholesale, and processing sectors. The limits to horizontal integration in farming, imposed by environmental factors that restrict cultivation within specific bounds, and the risks inherent in agricultural production caused by climatic and biological uncertainties (the very characteristics that have sheltered agriculture from corporate ownership and that limit farmers' ability to control their markets) have encouraged alternative forms of corporate control. Corporations, most often processing firms or feed companies, buy farms (ownership integration) or exert control through management agreements with farmers (contract integration), in order to minimise risk and ensure the profitability of the total corporate structure. The result is a tightening of the oligopolistic control of the market.

Both horizontal and vertical integration can help the retail and processing sectors increase their profits with advertising, the use of brand names, quality control, innovations and product development. However, increased physical efficiency in marketing through integration and concentration has reduced the public bidding for produce and so led to greater difficulty in establishing market prices. This is a particular problem when the wholesale market survives almost solely to clear poor-quality, non-standardised goods. As a result, competition at the processing and retail levels is now based on a range of marketing strategies, such as advertising and improved quality control, of which price setting is only one. Of those processing firms involved in agricultural marketing, most have extended vertical control either forward into retailing or backward towards primary production. The inherent risks in farming are such, however, that corporate management is exerted through market management and control of supplies, and not through the direct ownership of land. In the early 1970s, because of these risks, some large corporations in the United States pulled out of farming (*Agra Europe,* 1972). Only 0.4 per cent (as of 1981) of Canadian farms and 11.4 per cent (as of 1978) of those in the United States are owned by non-family corporations.

Farmers have responded to the full integration of agriculture into the commercial market system with a series of structural adjustments. Increased specialisation, larger farms, the application of science and technology to production, and the replacement of labour by capital-intensive techniques are all part of the changes that occur because of competition from corporate agribusiness in an open economy. However, the geographical and biological characteristics of agricultural production and the seasonal ebb and flow of farm labour requirements prevent a thoroughgoing extension of this process of structural adjustment. Moreover, farms remain small land-extensive businesses. The relatively limited scale of production and volume of output supplied by individual farmers constrain their ability to exploit the economic potential of labour specialisation or to establish market power by obtaining a regional monopoly of sales of a particular product. Similarly, the very risks inherent in farming through its continued dependence on fickle weather conditions and on other forces outside its control prevent the complete takeover of farming by corporate bodies.

Farm-Market Links

An increase in the price of a product is not usually sufficient to provoke an increase in agricultural production. Price incentives fail to increase commercial production where for example, as in many Soviet bloc countries, there is an adequate supply of consumer goods available for farmers to buy. Although poorly documented, factors such as taxation policy may have a similar dampening effect on farmers' output in the West, even where farm prices increase. In the Third World, the problem of persuading farmers to participate more fully in the commercial market has proved particularly intractable. Various reasons have been cited, including farmers' preference for leisure or increased consumption over an increased cash income, the inadequacy of consumer goods, and the high risk associated with new technology (Johnson, 1970).

In the past, the links between farmers and consumers were usually simple and direct. Local farmers could respond immediately to consumers' changing needs and preferences. In a modern economy this is not so. A new series of intermediaries, most of them urban-based, has sprung up. Processors, packers, wholesalers, distributors and retailers all contribute to the flow of information and produce between the two 'anchor' regions (farmers and consumers) at either end of the food chain. These intermediaries directly affect farm production patterns and consumer demands. This increased complexity requires close co-ordination among the different agents in the market system to ensure a rapid response to changes in market demand.

Supply-demand co-ordination based on price incentives is inadequate to meet the needs of a modern, industrial economy. The need for improved mechanisms to co-ordinate supply and demand has led to changing relationships between farmers and their markets, improved co-ordination of decision-making at all levels in the market system from farmer to consumers, and better-structured information flows between farmers, processors, wholesalers, retailers and consumers.

Modern technical requirements in the food processing industry call for the standardisation of raw materials, rigid quality controls, and the tight scheduling of supplies (Moore and Hussey, 1965). These concerns match the demands of the retail sector. However, modern markets are dynamic and consumers are fickle. Growth and competition within the food industry are increasingly based on the ability of processors and retailers to develop new products and to persuade consumers of their products' superiority in taste, price, quality, design, colour or convenience. New procurement policies adopted by processors, retailers and others in the food industry are designed to meet a wide range of corporate objectives, to contribute to higher profits, a larger market share, steady growth and an assured competitive edge on rival businesses. In all this, vertical controls play a predominant part.

Despite the pressures exerted by their market for regular bulk supplies of uniform quality, farmers remain constrained by biological limits that result in seasonal flows, small production lots and irregular quality controls. These constraints are frequently compounded by limited resources and an impaired information flow, so that farmers may remain only partially aware of market needs as the price received for their produce declines, and may be unable to respond effectively to this situation. Vertical integration offers an attractive means to overcome these constraints and to help match supply with market needs.

The strengthening of farm-market links has been largely through contract agreements between farmers, farm suppliers and processors. Using contracts, corporations control quality, volume and type of output. Contracts may even dictate the schedule of farm activities. The precise level of contract control varies regionally and among different types of farming. Control may be exerted simply through a written agreement. However, some contracts with livestock farmers, for example, require them to raise only livestock provided by the corporate partner and to use specified quantities and types of feed and antibiotics supplied by the contract firm. In agreement with farmers who grow crops, the contract may include the supply of seed, fertilisers, herbicides and pesticides. Contracts may allow the corporation to specify dates of planting and harvesting (which may be carried out by the corporation itself) and permit direct corporate supervision of animal management. In such situations the cost of farm inputs is usually deducted after the commodities have been delivered to the market.

Contracts *per se* are neither good nor bad. They can have wide-ranging social implications, but their precise impact depends on how risk is distributed and (if vertically integrated farms do indeed operate with greater efficiency) on how gains are shared. If a corporation buys a farm (ownership integration) the benefits of integration will go to the integrating firm, unless they are passed on to the consumer. Although ownership integration is not extensive, where it does exist it can put farmers at a disadvantage by reducing the size of the market open to them. In addition, independent firms may be squeezed out by the ability of larger integrated companies to survive for long periods on low profit margins in order to achieve long-term goals. Farmers who cannot get access to markets and who have no production alternatives must abandon their land.

With contract integration, farmers usually get some share of the gains, but the distribution of gains varies with the nature of the contract signed. Although group negotiation between farmers and processors can ensure a fair deal for both parties, farmers often have to negotiate contracts individually. Assured market outlets and income encourage farmers to specialise by providing them with an alternative to traditional mixed farming as a way of avoiding risk. Perhaps the most significant effect of contracts from an economic standpoint, is that they ensure a flow of capital and technology into farming. However, many contracts are short-term and the risks of agricultural production remain with the farmer.

A contract may offer a farmer a competitive edge by, for example, encouraging the use of more profitable production techniques. However, in the face of increased profits and the absence of any major barrier against new entrants into farming, increased output by more producers may erode this advantage. Small, spatially fixed, and bound by a contract, the farmer may ultimately be even more vulnerable than before.

Market Power

Market power may be narrowly defined as the ability to influence prices. However, market power in the modern world embraces a range of elements such as procurement policies, reaction to competition and market share. Consequently, attempts using only one index (commonly market share) to identify who has market power and how this power is exercised are seriously flawed (Parlby, *et al.*, 1976). Discussion of a shifting power focus and evidence of power abuse has tended to emphasise horizontal competition in specific markets with examples of overpricing, excess profits, or dubious business practices, and not the balance of forces between different sectors. Nonetheless, specialisation and concentration among retailers, wholesalers and processors have themselves promoted vertical linkages, not only as a deliberate strategy of risk avoidance, but as a means to assure supplies and market outlets, to control the cost of inputs, and to influence profit levels.

For the most part, in no single sector in the agriculture-food system does any one company or group of companies hold total sway. For certain items, such as sugar or canned soup, a few giant corporations may dominate. However, few foods are essential. Most commodities have ready substitutes (for example, butter can be replaced by margarine). The ability of consumers to alter their purchasing patterns when the price of a commodity increases should limit the extent to which any one corporation can make excessive profits. In addition, although one large corporation may dominate sales of any one product line, normally a few small firms also survive to offer some competition and consumer choice. Despite this, reliable statistical estimates show that consumers do have to pay higher prices where one company dominates any particular food sector (Parker and Connor, 1979).

Environmental factors constrain the extent to which a farmer can expand production of a particular crop and so limit any one farmer's ability to corner the market. These same environmental factors dictate the location and scale of processing plants. Canadian experience shows that although a few large retail chains may dominate regional markets, independent retail groups can compete effectively and expand to obtain an increased market share. Arguably, in all sectors of the food supply system — farmer, processor, wholesaler and retailer — small firms perform a distinctive and valuable economic role and at the same time, fill particular locational niches.

Small firms divert potential criticism from monopolies by meeting market needs that larger firms are unable or unwilling to meet. These may include long opening hours, home deliveries and the stocking of ethnic foods to satisfy local tastes. Likewise, small processing firms may produce specialised items or serve specific regional markets. Many small farmers earn good incomes by catering to markets that are prepared to pay a premium for high-quality goods unavailable elsewhere. Provision of these goods relieves consumer pressure for change on large firms.

The distinctive roles performed by firms of different sizes find geographical expression in the landscape. Location becomes an index of market power. The retail outlets of the large chains are located where consumer purchasing power is most concentrated. Independent retail stores serve inner city areas and small communities. Major processing firms are located near key production areas and strictly delimited, concentric zones of production are found around processing plants (Coppock, 1971). In particular, zones of contract farming are found close to processing firms. Although large-scale farming can survive independently of contracts and at some distance from major urban markets, small farms without contracts and with few market options may decline in these areas. However, certain regions that could support specific types of farming do not do so because market facilities are unavailable.

Market power cannot be directly equated with a firm's share of a specific market. Rather, power in the market system is related to flexibility, mobility and location. Within any one sector there may be some concentration of power, but any assessment of that power must include not only market concentration, but vertical controls. The failure to recognize vertical controls as an important measure of market power and to focus solely on market concentration has encouraged attempts to promote a power balance in the market system by increasing concentration in other sectors — particularly in farming.

Countervailing Power

Farmers' sense of vulnerability to what they perceive as the misuse of market power by the corporate sector has led to repeated calls for government intervention on their behalf. Understandably, farmers' accusations of corporate exploitation are most marked during periods of low prices for farm produce. Government-sponsored intervention in the market system has taken two main forms — co-operatives and marketing boards (see Tarrant, 1974). Both forms of intervention are found in industrialised countries and in the Third World. However, for the most part, market intervention in the Third World is part of a colonial legacy designed and maintained to ensure low prices for urban consumers, not as elsewhere, to boost farmers' power (Bates, 1983).

Co-operatives

Farm co-operatives are voluntary associations of individual producers who combine to increase their buying or selling power. Although not government controlled they are often backed by specific legislative policies and by government funds. The value of co-operatives in obtaining economies of scale in the handling of farm produce and the purchase of capital equipment or farm supplies is widely accepted. Their marketing power, however, is severely limited. Galbraith has attributed this to a series of structural flaws (1952). Co-operatives are loose associations of individuals and rarely include all producers of a given product. The co-operative controls neither the members' level of production nor their release of goods onto the market. Whereas a strong bargaining position requires the ability to wait and to withhold goods from the market, a co-operative has no control over non-members, who can sell what they please when they please. In practice, a co-operative cannot control even its own members who, if prices do rise, are tempted to break away and sell all they can produce at the going rate.

In certain parts of the world, in particular some of the small liberal democracies of northern and western Europe, co-operatives have increased the power of farmers in the market system by group negotiations for higher farm prices. However, such success seems to depend on favourable social conditions and the willingness of farmers to be bound by the rules of the co-operative. Co-operatives have not proved as successful elsewhere. In North America and Africa, for example, the power of co-operatives has been curtailed by such factors as the limited volume of production they control, lack of producer loyalty, poor managerial ability and weak financing.

Moves to overcome these problems have only created new problems and increased the dissatisfaction of co-operative members. Like large corporations in which management perpetuates itself, the management of large co-operatives may become distant from the members and it may be difficult to maintain membership control of management decisions. Members may distrust one another and distrust management, but remain bound to the co-operative as there is no alternative market outlet for their produce (Williams, 1980). Consequently, despite their success in certain areas, marketing co-operatives often seem inappropriate to the exisiting economic system (Allen, 1975). It is this inadequacy that has prompted the introduction of producer-controlled marketing boards.

Marketing Boards

In marked contrast to co-operatives, membership on marketing boards is compulsory. This is essential if the marketing board is to control output and prices. Although their function may vary from promoting farm products and keeping farmers abreast of market demands to quality control and price setting, criticism has centred on their use and abuse of power — particularly the power to regulate supply and price.

The primary aim of marketing boards with supply management powers is to ensure stable and equitable returns for farmers and good quality food at reasonable prices for consumers. Marketing boards vary widely in their form and structure and in the mechanisms they use to achieve their goals. In Canada, boards with the power of supply management cover sectors such as poultry, eggs and dairy products. These boards rely primarily on production quotas to achieve their objectives. When supply management is initiated, quotas are allocated free to established producers on the basis of their previous output. Over time, quotas are increased or decreased as necessary and guidelines are established for allocating new quotas and redistributing quotas that become available when an established producer ceases production. Whatever guidelines are adopted, the quotas invariably acquire a cash value. In some cases, they are sold openly on the market. Where this is forbidden, the value of quotas may be incorporated into the cash valye of a farm. The net result is an increase in the capital costs associated with farming and an increased barrier to new entrants to the industry.

In so far as an increase in the price received by farmers reflects an increase in their market power, supply management as applied through the market board structure is a proven success. However, this one measure of market power is inconclusive. The use of supply management to stabilise prices has increased corporate control of farming and speeded up the rate of decline in farm numbers, as farmers granted quota rights can stop production and sell these rights. For these producers, quotas amount to a free cash gift. The net result of supply management, therefore, may be to increase farmers' market power, but this power devolves into fewer hands. Where supply management increases corporate control of farming, the improvement of farmers' power is illusory. The final result may be an acceleration of pre-existing trends towards the industrialisation of agriculture.

Government Marketing Policies and National Development Plans

The agricultural market system is commonly the scapegoat for much that it is not directly responsible for (Allen, 1959). Government intervention in agricultural commodity markets occurs throughout the world. In western market-oriented economies the objectives of market intervention include stability of prices and quantities of output, higher farm incomes, and the maintenance or increase in the efficiency of resource use in farming (Heidhues, 1976). These often conflicting aims have their counterpart in developing countries where low food prices for urban consumers are maintained at the expense of a failure to raise total food output (Bates, 1983; Priebe and Hankel, 1981). This situation is often compounded by serious losses due to inadequate storage and transportation capacity (Greeley, 1982). In the Soviet Union and Eastern Europe, similar pricing policies and post-harvest management problems have also stifled agricultural growth (Symons, 1942; Zeman, 1978).

Government intervention in agricultural markets has a direct impact on market prices. Geographical theory suggests that these influence the pattern of land use. In addition to price, a range of interventionist strategies including control of entry into specific types of farming (as through the use of production quotas), transportation subsidies, trade barriers and land zoning regulations all impinge on the rural landscape (Morgan and Munton, 1971). However, few geographers have examined the specific impact of government marketing policies on agricultural land use, although a small number of studies recognize the importance of a range of government policies (including marketing policy) on patterns of agricultural production (Fielding, 1964, 1965), trade (Tarrant, 1980), and development (Wanmali, 1980).

Frequently, agricultural marketing is only one part of a comprehensive national agricultural plan and such plans are themselves increasingly designed to encompass the total agriculture-food system within which the market is viewed as a planning tool (Adeyemo, 1984; Binns, 1982; Bloomfield, 1983; Canada, 1977; Epstein, 1982; Famoriyo, 1978; Kaynak, 1981; Mittendorf, Barker and Schneider, 1977). Designed to meet specific agricultural objectives such as increased food self-sufficiency or low food prices, or to maintain the size of the farm population, these national plans often also aim to use agricultural growth to bolster other sectors of the economy, such as manufacturing, to influence social equity through the manipulation of prices and incomes, and to ensure the stability of the current political regime. Despite the importance of government intervention (in Canada, for example, more than half of the total agricultural output is under some form of direct government control) institutional factors remain seriously neglected by agricultural geographers and no suitable analytical model has evolved.

Conclusions

The difficulty of disentangling the market issue from broader agricultural policy concerns; the inherent complexity of market relationships, and their wide-ranging impact, all pose serious difficulties for geographers anxious to explore this important yet neglected field. The von Thünen model is based on the assumption of a relatively simple and direct price relationship between farmers and their markets. This is inadequate for the analysis of modern market relationships where a number of intermediaries separate farmers and consumers and where price is only one of many mechanisms used to co-ordinate supply and demand.

The need for a new conceptual framework for market analysis and attempts to provide one are recognised both in economics (Shaffer, 1980) and in geography (Smith, 1984). Both authors have identified the ideas proposed by Hirschman (1970) as the basis for an organisational structure which could describe the existing systems for production and distribution of food, help identify the economic and institutional forces that influence this system, and help design market policies that would serve the multiplicity of objectives held by different sectors of the community. Thus in the modern economy, where there are wide areas of monopoly and oligopolistic competition, it is futile to place continued reliance on a simple balance of economic forces resting on assumptions of comparative advantage and perfect competition. Hirschman identifies 'exit' and 'voice' as two primary mechanisms of equal weight and complementary character used to alter market performance and to limit decline. (Exit refers to the basic economic mechanism inherent in models of price competition, the consumer's ability or willingness to 'vote with his feet', to refuse to buy and to switch his purchasing power elsewhere. Voice is the alternative 'political' option, the vocal protest directed at management, government, or anyone else who will listen.) Hirschman also includes a theory of loyalty, not as an additional mechanism, but to offer an effective means to explain the use of the two alternative mechanisms and to interpret their impact.

Grigg (1983) has questioned the view that the absence of theory thwarts progress in agricultural geography and suggests that possibly the relative paucity of agricultural geographers is an even more important stumbling block. Nevertheless it is clear that a good conceptual framework could stimulate research interest and focus attention on key issues. Its absence should not be used to cover a sterile retreat to more comfortable, established, research themes.

References

Adeyemo, R. (1984) 'The food marketing system: implications of the Green Revolution Programme in Nigeria', *Agricultural Systems, 14*, 143–57

Allen, G.R. (1959) *Agricultural Marketing Policies*, Basil Blackwell, Oxford

Allen, G.R. (1975) 'Changes in the relationships between agriculture, the food industry and trade: markets and marketing', *European Review of Agricultural Economics, 2*, 433–57

Agra Europe (1972) 12 April, cl/1

Bates, R.H. (1983) 'Patterns of market intervention in agrarian Africa', *Food Policy, 8*, 197–304

Binns, J.A. (1982) 'Agricultural change in Sierra Leone', *Geography, 67*, 113–25

Bloomfield, I.C. (1983) 'National food strategies and food policy reform', *Food Policy, 8*, 287–96

Bowler, I.R. (1981) 'Some characteristics of an innovative form of agricultural marketing, *Area, 13*, 307–14

Bowler, I.R. (1984) 'Agricultural geography', *Progress in Human Geography, 8*, 255–62

Breimyer, M.F. (1965) *Individual Freedom and the Economic Organization of Agriculture*, University of Illinois Press, Urbana, Illinois

Brown, L.R. (1984) 'A crisis of many dimensions: putting food on the world's table', *Environment, 26*, 15–20, 38–43

Canada, Department of Agriculture and Department of Consumer and Corporate Affairs (1977) *A Food Strategy for Canada*, Ottawa

Chisholm, M. (1962) *Rural Settlement and Land Use*, Hutchinson, London

Coppock, J.T. (1971) *An Agricultural Geography of Great Britain*, Bell, London

Cracknell, M.P. (1980) 'Multinational food companies and agriculture', *World Agriculture, 29*, 16–19

Cromley, R.G. (1980) 'The isolated state: an agricultural location game', *Journal of Geography, 79*, 230–34

Ehrensaft, P. (1983) 'The industrial organization of modern agriculture', *Canadian Journal of Agricultural Economics, 31*, 122–33

Epstein, T.S. (1982) *Urban Food Marketing and Third World Development*, Croom Helm, London

Famoriyo, S. (1978) 'Food production policies in Nigeria', *Food Policy, 3*, 50–8

Fielding, G.J. (1964) 'The Los Angeles milkshed: a study of the political factor in agriculture', *Geographical Review, 54*, 1–12

Fielding, G.J. (1965) 'The role of government in New Zealand wheat growing', *Annals of the Association of American Geographers, 55*, 87–97

Found, W.C. (1971) *A Theoretical Approach to Rural Land Use Patterns*, Macmillan, Toronto

Galbraith, J.K. (1952) *American Capitalism: The Concept of Countervailing Power*, Houghton Mifflin, Boston

Greeley, M. (1982) 'Pinpointing post-harvest food losses', *Ceres, 15*, 30–7

Gregor, H.F. (1970) *Geography of Agriculture: Themes in Research*, Prentice-Hall, Englewood Cliffs, New Jersey

Grigg, D. (1983) 'Agricultural geography — progress report', *Progress in Human Geography, 7*, 256–9

Hebden, R.E. (1984) 'The 1982 Soviet food programme', *Geography, 69*, 62–3

Heflebower, R.B. (1957) 'Mass distribution: a phase of bilateral oligopoly or of competition?' *American Economic Review, 47*, 274–85

Heidhues, T. (1976) 'Price and market policy for agriculture', *Food Policy, 11*, 116–29

Hirschman, A.O. (1970) *Exit, Voice and Loyalty*, Harvard University Press, Cambridge, Mass.

Huff, J.O. (1981) 'Richman — poorman in von Thünen's isolated state', *Economic Geography, 57*, 127–33

Johnson, E.A.J. (1970) *The Organization of Space in Developing Countries*, Harvard University Press, Cambridge, Mass.

Jones, D.W. (1982) 'Location and land tenure', *Annals of the Association of American Geographers, 72*, 332–46

Jumper, S.R. (1974) 'Wholesale marketing of fresh vegetables', *Annals of the Association of American Geographers, 64*, 387–96

Kaynak, E. (1981) 'Food distribution systems', *Food Policy, 6*, 78–90

Luqmani, M. and Quraeshi, Z. (1984) 'Planning for market coordination in LDCs: the role of channel participants in improving food distribution systems', *Food Policy, 9*, 121–30

Mallen, B. (1976) *A Preliminary Paper on the Levels, Causes and Effects of Economic Concentration in the Canadian Retail Food Trade: A Study of Supermarket Power*, Reference Paper 6, Canada, Food Prices Review Board, Ottawa

Mittendorf, H.J. and Abbot, J. (1979) 'Provisioning the urban poor: the challenge in food marketing systems', *Ceres, 12*, 26–32

Mittendorf, H.J., Baker, E.J. and Schneider, H. (1977) *Critical issues on food marketing systems in developing countries*, Organisation for Economic Co-operation and Development (OECD), Paris

Moore, H.L. and Hussey, G. (1965) 'Economic implications of market orientation', *Journal of Farm Economics, 47*, 421–7

Morgan, W.B. and Munton, R.J.C. (1971) *Agricultural Geography*, Methuen, London

Norton, W. (1979) 'The relevance of von Thünen theory to historical and evolutionary analysis of agricultural land use', *Journal of Agricultural Economics, 30*, 39–47

Parker, R.C. and Connor, J.M. (1979) 'Estimates of consumer loss due to monopoly in the US food-manufacturing industries', *American Journal of Agricultural Economics, 61*, 627–39

Parlby, G., Famure, O., Faminow, M., and Hawkins, M.H. (1976) 'A critical review of the Mallen Report', *Canadian Journal of Agricultural Economics, 24,* 40–9

Priebe, H. and Hankel, W. (1981) 'Agricultural policy in developing countries', *Intereconomics, 16,* 31–6

Shaffer, J.D. (1980) 'Food system organization and performance: towards a conceptual framework', *American Journal of Agricultural Economics, 62,* 310–18

Shakow, D. (1981) 'The municipal farmer's market as an urban service', *Economic Geography, 57,* 68–77

Skinner, G.W. (1967) 'Marketing and structural change in rural China' in J.M. Potter, M.N. Day, and G.M. Foster, (eds.), *Peasant Society: a Reader,* Little, Brown, Boston, pp. 63–98

Smith, R.H.T. (1979) 'Periodic market-places and periodic marketing: review and prospect — I' *Progress in Human Geography, 3,* 471–505

Smith, R.H.T. (1980) 'Periodic market-places and periodic marketing: review and prospect — II', *Progress in Human Geography, 4,* 1–31

Smith, W. (1974) 'Market-farm linkages and land use change: a Quebec case study', *Cahiers de Géographie de Québec, 18,* 297–315

Smith, W. (1984) 'The "vortex model" and the changing agricultural landscape of Quebec', *Canadian Geographer, 28,* 358–72

Symons, L. (1942) *Russian Agriculture: A Geographical Survey,* Westview, Boulder, Colorado

Tarrant, J.R. (1974) *Agricultural Geography,* David and Charles, Newton Abbot

Tarrant, J.R. (1980) 'Agricultural trade within the European Community', *Area, 12,* 37–42

United States Department of Agriculture (1981) *A Time to Choose: Summary Report on the Structure of Agriculture,* USDA, Washington, DC

Visser, S. (1980) 'Technological change and the spatial structure of agriculture', *Human Geography, 56,* 311–9

von Thünen, J.H. (1966) *von Thünen's Isolated State,* translated by C.M. Wartenburg, with an introduction by P. Hall, Oxford University Press, Oxford

Wallace, A.I. and Smith, W. (in press) 'Agribusiness in North America' in B.W. Ilbery and M. Healey (eds.), *Industrialisation of the Countryside,* Geo Books, Norwich

Wanmali, S. (1980) 'The regulated and periodic markets and rural development in India', *Transactions of the Institute of British Geographers, 5,* 466–86

Williams, R.E. (1980) 'Milk marketing in a European framework', *Journal of Agricultural Economics, 31,* 311, 320

Zeman, Z. (1978) 'Economic planning of Eastern Europe and the USSR: the role of agriculture', *Food Policy, 3,* 127–35

10 THE WORLD FOOD PROBLEM

D.B. Grigg

Over the last half century there has been much concern over the extent of hunger, and particularly that in Afro-Asia and Latin America. In the 1930s the League of Nations stated that two-thirds of the world's population were undernourished (League of Nations, 1936) a figure repeated by Sir John Boyd Orr, (Boyd Orr, 1950) the first director of the Food and Agriculture Organization, in 1950. More recent estimates are lower, but more variable. Thus (Table 10.1) FAO estimated that in 1980 436 million people in the developing countries (excluding China) were undernourished, 19 per cent of the total; another authority however has put this figure as high as, in 1975, 1373 million, 71 per cent of the population of the developing countries. A recent critical review of this latter figure however, argues that it should be much lower. Lipton believes that only 10–20 per cent of the population of most developing countries have diets that adversely affect their health, or physical and mental development. This yields a total, in 1980, of between 219 and 438 million, excluding China (Lipton, 1983), similar to the FAO figure.

Table 10.1: Recent Estimates of the Extent of Undernutrition (Millions)

Year	Far East	Near East	Africa	Latin America	Total
1972–4	297	20	83	46	446
1975	924	94	243	112	1,373
1980	303	19	72	41	436
		(% of population)			
1972–4	29	16	28	15	25
1975	82	51	77	36	71
1980	23	9	20	11	19

Note: All estimates exclude China.
Sources: 1972–74: FAO, 1977; 1975; S. Reutlinger and M. Selowsky, *Malnutrition and poverty: magnitude and policy options*, World Bank Staff Occasional Papers, no. 23, 1976, p. 31; 1980: FAO, *The State of Food and Agriculture 1981*, Rome, 1982, p. 75.

The Problem of Measurement

It is not surprising that there is such a variation in the estimates of those with poor diets. An undernourished man is one who ingests an insufficient amount of energy to maintain his Basal Metabolic Rate and to carry out his normal work. If undernourished his weight will fall and so too will his capacity to work. Although surveys of body weights exist they are too few to give an indication of undernutrition at national, regional or world levels. Malnutrition, due to an inadequate intake of protein, and/or a range of vitamins, causes specific diseases which can only be established by medical diagnosis, and such surveys cover only small fractions of the population of the developing countries. Counting those suffering from malnutrition is complicated by two futher facts. First, young children in the developing countries suffer from numerous diseases of the stomach and intestines, and the symptoms of malnutrition may be due not to an inadequate diet but their inability to digest the food they eat. (Mata *et al.*, 1977) Second, the severity of protein calorie malnutrition — and indeed other deficiency diseases — may vary greatly. Thus in India it was estimated that only 1.2 per cent of children suffered from kwashiorkor which, untreated, can cause death, but 80 per cent showed signs of reduced growth for their age. (Gopalan, 1975). Thus the numbers of malnourished may vary widely according to the degree of severity used as a criterion.

Food Balance Sheets

In the event most estimates of the extent of undernutrition depend upon food balance sheets prepared by FAO for most countries in the world. The first step is to estimate the total food output in a year, and convert this into the calorific equivalent; from this is deducted the seed needed for the next harvest, the food crops used for industrial purposes, and those exported. Added on are the stocks carried over from a preceding year, and imports of food. The data are then expressed as calories *per capita* per day. The average for the developed countries in 1978–80 was 3,407, for the developing countries 2,328, ranging from 3,938 for Belgium to 1,729 for Ethiopia (FAO, 1983).

These figures must then be compared with estimated requirements. However, measuring the minimum necessary calorie intake — which is assumed also to provide sufficient protein and vitamins — is exceedingly difficult. FAO and the World Health Organisation have published minimum requirements for different ages, sexes and levels of activity, (FAO/WHO 1973), but these have been subject to much criticism; it is generally agreed they overstate minimum requirements and the figures are currently being revised. Using these estimates, together with national data on age structure and the sex ratio, FAO have estimated the minimum *national* requirements for all countries (FAO, 1977). These ranged from 2,160 calories in Indonesia to 2,690 calories in Sweden. The difference is due to the much greater proportion of infants and children, with lower calorie requirements, in the developing countries than in the developed, with their higher proportions of adults.

The Distribution of Undernutrition

These data have been used in various ways to show spatial variations in undernutrition. If the available food supplies were distributed solely according to need as determined by weight, sex and age, then a country should have no problems of undernutrition if available food supplies are 100 per cent or more of national requirements (Figure 10.1). Thus, as expected, there are on these grounds no nutritional problems in any of the developed countries; but substantial parts of the developing world also have available food supplies in excess of 100 per cent — China, all of South-east Asia except Laos, Vietnam and Kampuchea, nearly all of North Africa and the Middle East, most of South America, and Mexico and Cuba. Indeed in 1980 two-thirds of the population of the developing world lived in countries where food supplies exceeded requirements. Those countries where available food supplies are less than requirements are: first, most of the countries of tropical Africa; second, India, Bangladesh and Nepal; third, Vietnam, Kampuchea and Laos; and fourth, a number of countries in the Andes and Central America. As it is often said that the problem of undernutrition is not one of production but of distribution, it should be noted that in these countries, which contain one-third of the population of the developing world, home production, stocks and imports were *insufficient* in 1978–80 to provide all the population with an adequate diet even if the available food supplies had been distributed according to need alone. They can only solve their problems either by increasing output or by increasing imports.

However food supplies are not distributed according to nutritional need in any country. The principal reason for differences in consumption between individuals or groups within a country is income, or in the case of subsistence farmers, the amount of land they have. In the

Figure 10.1: Available Calories *Per Capita* per Day as a Percentage of National Minimum Requirements

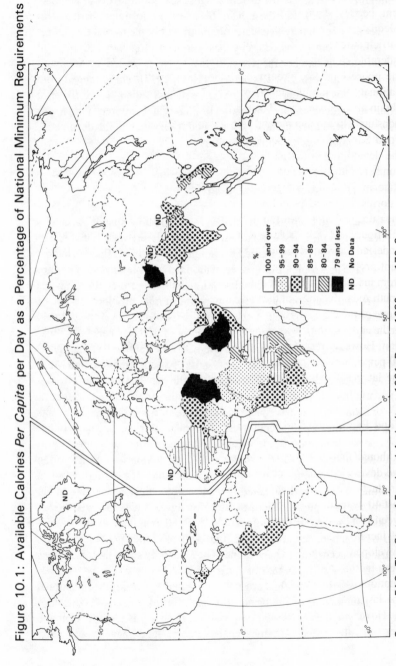

%

100 and over
95-99
90-94
85-89
80-84
79 and less
ND   No Data

Source: FAO, *The State of Food and Agriculture 1981*, Rome, 1982, pp. 172-3.

developed countries undernutrition is not, for the most part, a major problem, because even the lowest income groups have sufficient to buy an adequate diet. In developing countries the upper income groups have sufficient money to buy an adequate diet and indeed a diet with a considerable element of expensive animal foodstuffs. On the other hand, the poorest income groups, including those in countries which have national food supplies in excess of national requirements, are unable to obtain an adequate diet. There have been numerous studies of the relationship between income and calorific consumption; in the Fourth World Food Survey, making some assumptions about income distribution in the developing countries, estimates were made of the proportion and number of the population who obtained less than 1.2 x the Basal Metabolic Rate in 1972–74, a figure that varied from 1,486 calories to 1,631 calories, and was much lower than the minimum requirements used to measure *national* food requirements. (Table 10.1) FAO have updated this figure to 1980 (Table 10.1) and the approximate distribution of those underfed can be calculated for the same year (Figure 10.2). A significant proportion of the population of most developing countries suffer from inadequate diets, but in absolute terms it is in Asia, and particularly South Asia, that most of the undernourished are to be found (Table 10.1). None of the calculations in Table 10.1 includes estimates for China, and in the early 1970s it was assumed that rationing had eliminated malnutrition. However recent statements by the Chinese government suggest that 10 per cent of the population — 100 million — had inadequate diets in the late 1970s (Lardy, 1982).

Food Production and Population Growth

Although there are very considerable differences in productivity between the developed and the developing world, this does not mean that food production in Afro-Asia and Latin America is static. Indeed over the last 40 years there has been a remarkable increase in food output. Between 1950 and 1980 world food output more than doubled, and the rate of increase has been greater in the developing countries than in the developed. (Table 10.2). In all the developing regions food output has been increasing at 2 per cent per annum for three decades, with the exception of Africa in the 1970s where, although output has increased in the last decade, it has done so at a diminished rate.

However in the same period world population has increased by 76 per cent, that of the developed world by 36 per cent, and that of the

Figure 10.2: Percentage of the Population Receiving Less than 1.2 × Basal Metabolic Rate, 1978–70

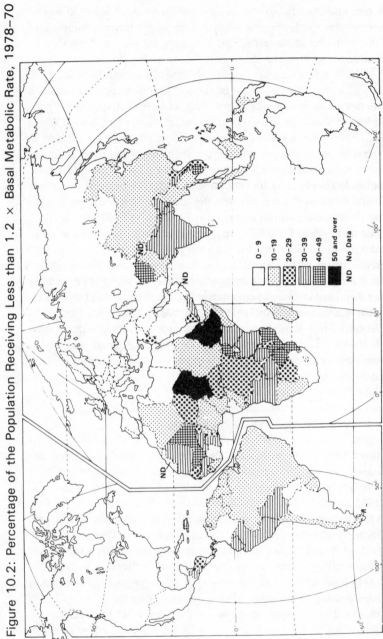

0 – 9
10 – 19
20 – 29
30 – 39
40 – 49
50 and over
ND   No Data

Source: FAO, *Production Yearbook 1982*, Rome, 1983; FAO 1977.

Table 10.2: Rate of Increase in Food Output, 1952–1980 (% per annum)

	1952-4 to 1959-61	1961-1970	1971 to 1980
Africa	2.1	2.7	1.8
Far East	3.4	3.5	3.6
Latin America	3.1	3.5	3.8
Near East	3.3	3.0	3.5
Asian CPE	n.a.	2.7	3.2
All Developing	3.1	3.1	3.3
All Developed	3.0	2.4	1.9
World	3.1	2.7	2.5

developing countries by 95 per cent. The great increase in the population of the developing regions has been due to an abrupt decline in mortality combined with little change in fertility until the 1960s. Since the end of that decade there has been some decline in fertility in parts of the developing world, most noticeably in China. Hence, the rate of overall increase in population has slowed, for since the 1960s the rate of decline in the death rate has also diminished. (Gwatkins, 1980). To these generalisations Africa is a major exception, for there has been no change in fertility levels: by the late 1970s the population of tropical Africa was increasing faster than at any time in the past.

In the 1950s many writers argued that population growth would inevitably outrun food production, and food supply *per capita* fall, with catastrophic results. Yet as noted earlier, food output has grown very rapidly. The relationship between food output and population growth varies according to the scale and the period considered.

(1) At the world level food output has grown more rapidly than population (Table 10.3). Simon has estimated that world food output *per capita* increased by 28 per cent between 1948–52 and 1976 (Simon, 1980); output per head rose a further 3.5 per cent from 1974–6 to 1982 (FAO, 1983).

(2) At the regional level food output *per capita* has risen strikingly in the developed world because of a combination of rapid increase in food output and a comparitively slow rate of population increase (Table 10.3).

(3) In the developing regions increases in food output *per capita* have been much more modest, and in Africa there was a small decline in the 1950s, and a very alarming decline in the 1970s (Table 10.3; Figure 10.3).

(4) However if the food ouput *per capita* is considered at the country level, then the failure to keep up with population growth seems more widespread. Data on food output *per capita* at the country level are only available since 1961–5, and not as yet in a continuous series.[1] However 50 countries had a decline in food ouptut *per capita* from 1969–71 to 1978–80, 33 of these in Africa, and 22 had a decline in both 1961–5 to 1969–71 and 1969–71 to 1978–80. However many of these countries still had food supplies above national requirements in 1978–80; Chile, Uruguay, Saudi Arabia and Algeria are examples. Those countries which both lacked adequate national food supplies and had a decline in food output *per capita* for one or two decades are shown in Figure 10.4. Clearly Africa has the most widespread problems.

Table 10.3: Rate of Increase in Food Output *Per Capita*, 1952–1980 (% per annum)

	1952–4 to 1959–61	1961–1970	1971–1980
Africa	−0.2	0.1	−1.2
Far East	1.1	0.9	0.9
Latin America	0.3	0.8	1.2
Near East	0.8	0.3	0.6
Asian CPE	n.a.	0.9	1.6
All developing	0.7	0.7	1.0
All developed	1.7	1.4	1.1
World	1.1	0.8	0.6

Sources: FAO, *World Agriculture: the last quarter century,* Rome 1970, p. 9; *The State of Food and Agriculture 1981,* Rome, 1982, pp. 5–6; *The Fourth World Food Survey,* Rome, 1977, p. 4.

Thus in some countries food output has failed to keep up with food production in the last two decades. It does not follow from this that population growth is the sole cause or indeed always the major cause of undernutrition at present. In particular it is *not* the rapid post-war population growth that has given rise to spatial inequalities in food supply; in the 1930s the present pattern of calories available *per capita* already existed, with a marked difference between the developed countries and the developing (Bennett, 1941; FAO, 1952; Grigg, 1982). In spite of the great increase in population in the developing countries since then, there are few if any countries where food consumption *per capita* is now lower than it was in 1930 or in 1950. Conversely it is difficult to deny that population growth has had a serious effect upon the extent of under-nutrition. In many rural areas population growth has reduced the

Figure 10.3: Changes in Food Output *Per Capita*
1950–1980, by Developing Countries

Source: U.S. Department of Agriculture, *World Indices of Agricultural Food Production*, Washington, D.C. 1981.

size of farms, increased the extent of rural unemployment and underemployment, and led to falling real incomes for some of the rural population. This has been compounded, in some rural areas in the developing world, by the adoption of machinery, the amalgamation of farms and the increase in landlessness. A substantial proportion of the rural populations of the developing world have migrated to the towns since 1950; but this addition to the rapid *natural* increase in urban areas has made it difficult to create employment opportunities. Hence in spite of the rapid *national* economic growth in all developing countries since the 1950s, poverty remains a formidable problem, and the major cause of undernutrition.

Figure 10.4: Countries with Available Calories *Per Capita* Below National Minimum Requirements and which had Experienced a Decline in Food Output *Per Capita* 1969–71 to 1978–80 or 1961–5 to 1978–80

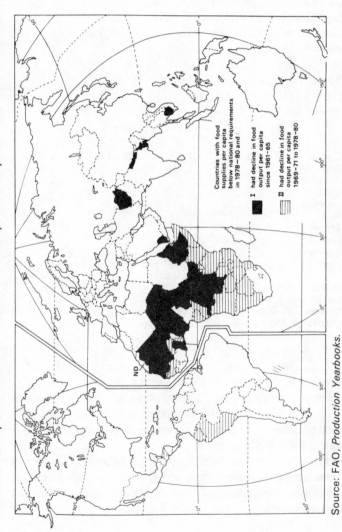

Source: FAO, *Production Yearbooks*.

Changes in the Extent of Hunger

If it is difficult to measure the present extent of hunger it is even more difficult to measure changes over time. Available food supplies *per capita* — which include imports — have increased in all the major regions since 1950, (Table 10.4) but this does not show how the total supplies are distributed amongst the population. However if the proportions and numbers receiving less than 1.2 x the Basal Metabolic Rate are calculated for 1950, 1960, 1970 and 1980 it can be seen that the *proportion* of the population of the developing world undernourished has declined, from 34 per cent to 17 per cent. The absolute numbers increased between 1948–50 and 1961–5, but have since declined (Table 10.5).

Table 10.4: Average Available Food Supply *Per Capita* per Day, 1950–1980 (calories)

	1950	1961–55	1969–71	1978–80	1950–80 % change
Europe	2,689[a]	3,420	3,339	3,477	29.3
North America	3,131	3,492	3,467	3,624	15.7
Oceania[b]	3,176	3,432	3,360	3,257	2.5
USSR	3,020	3,542	3,388	3,486	12.2
Developed	2,878	3,471	3,382	3,486	21.1
Asia[d]	1,924	2,068	2,192	2,326	20.9
Africa[e]	2,020[c]	2,165	2,276	2,311	14.4
Latin America	2,376	2,413	2,531	2,591	9.0
Developing	1,977	2,115	2,239	2,350	18.9
World	2,253	2,494	2,537	2,617	16.2

Notes: a. No data for Eastern Europe: assumed to be same as Western Europe for aggregated figures.
b. Australia and New Zealand only.
c. Data for 44% of population only; other countries assumed to have same average.
d. Includes China, Japan, Israel
e. All Africa, including South Africa
Source: FAO, *Second World Food Survey,* Rome, 1952; *Production Yearbook 1975,* Rome, 1977; *Production Yearbook 1982,* Vol. 36, Rome 1983.

It is widely agreed that the primary cause of undernutrition is poverty; families with low incomes cannot afford an adequate diet, either owing to low wages or — in the case of farmers — inadequate land. Thus the mere fact that a country has adequate minimum requirements does not mean that there is no undernutrition. This is a function of the distribution of income within the country. Historically Europe's nutritional

Table 10.5: Numbers Receiving less than 1.2 × Basal
Metabolic Rate 1950–1980 (Millions)

	1948–50	*1961–5*	*1972–4*	*1978–80*
Africa	60	94	83	72
Latin America	46	54	46	41
Asia	444	502	397	421
	550	650	526	534
	As % of total population			
Developing countries	34	29	20	17
All countries	23	21	13	12
	includes estimates for China			

Source: D.B. Grigg, *The world food problem 1950–1980* (forthcoming).

development provides some interesting parallels with the developing countries today. About 1800 few West European countries produced sufficient calories *per capita* to provide all with an adequate diet; in Germany and France availability *per capita* was about 2,000 calories, comparable with India today. However in the nineteenth century food output grew rapidly and, in spite of the rapid population growth, supplies reached 3,000 calories *per capita* per day by the later nineteenth century. (Figure 10.5). Most of this increase was in vegetable calories, mainly cereals and potatoes. Indeed in 1880 70 per cent of the French national consumption of calories was still provided by cereals and potatoes. (Figure 10.6). However from the 1890s whilst total calorie intake did not greatly increase (Figure 10.7), the proportion derived from animal products did; in Western Europe by the 1960s about one-third of all calorie consumption came from animal foodstuffs. After the mid-nineteenth century there was little evidence of undernutrition in Europe, but malnutrition remained widespread because, although national food availability was adequate, income distribution was such that the poorest groups were unable to purchase an adequate diet. As late as the 1930s a League of Nations report stated that there was no country in Europe free from malnutrition; in Britain in the 1930s surveys estimated that between one-third and one-half of the population had incomes too low to provide a good diet. (Boyd Orr, 1937; Le Gros Clark and Titmus, 1939). Only in the period since the end of the Second World War have the lowest income groups has been able to afford diets that have largely eliminated malnutrition.

In the developing world today there are many countries — containing one-third of the population — where total supplies are less than

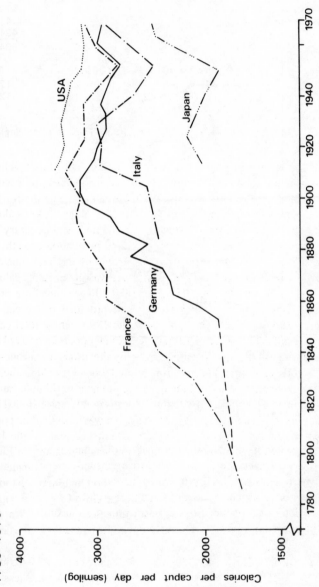

Figure 10.5: Changes in the Available National Calorie Supply *Per Capita* per Day, Selected Countries, 1780–1970

Source: A. Weber and E. Weber, 'The structure of world protein consumption and future nitrogen requirements', *European Review of Agricultural Economics*, 1974–5, 2, 169–192.

Figure 10.6: France, 1780–1970; (a) Changes in the
Percentage of All Calories Derived from Bread and Potatoes;
(b) Animal Protein as a Percentage of All Protein

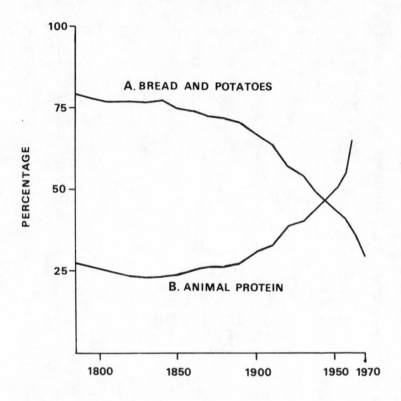

Source: J.C. Toutain, *La Consommation alimentaire en France de 1789 à
1964*, Paris, 1971.

national requirements and the elimination of hunger requires not only
an increase in incomes, but an increase in food output *per capita*.
Elsewhere food supplies are in excess of requirements — although in
many countries, such as India or China, only just in excess — and the
elimination of hunger requires an improvement in income if the problem of undernutrition is to be overcome.

Figure 10.7: France 1780–1962: Total Supply of Calories, Supplies from Vegetable Sources Alone and Supplies from Animal Products Alone

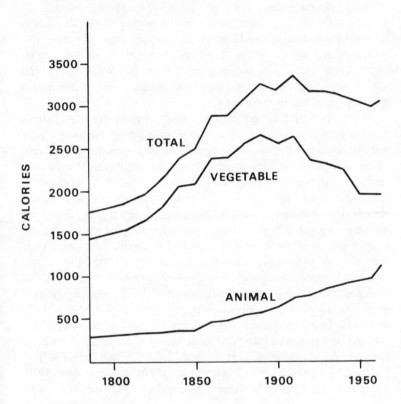

Source: As for Figure 10.6.

Income Growth and Income Distribution

Over the last 30 years the national wealth of nearly all developing countries has grown more rapidly than population; real Gross Domestic Product *per capita* in all the developing countries rose by 134 per cent between 1950 and 1975 (Morawetz, 1977). This remarkable advance needs some qualification. First, in spite of this improvement, many developing countries, particularly in Africa and parts of Asia, remain remarkably poor. Second there have been marked regional variations in the rate of growth. In some of the OPEC countries oil exports have given Gross Domestic Products *per capita* comparable to those of the developed world, whilst in Brazil, Mexico, Korea and Taiwan there has been considerable industrial growth. There has thus been a growing differentiation between the countries of the developing world, so that it is no longer possible to divide the world sharply into two blocs, the developed and the developing. The Gross Domestic Product *per capita* in the richer developing countries is many times that in the poorest, and some countries, particularly the oil exporters, can now afford to import not only grain but livestock products.

Third, although there has been a marked increase in national Gross Domestic Product *per capita* in nearly all developing countries, it does not follow that all sectors of the population have benefited. First, part of the Gross Domestic Production is used for investment and not consumption. Second, not all classes have experienced equal increases in their incomes. A large and tendentious literature has appeared on this topic in the last fifteen years. It has been argued, first that there has been an increasing gap between the rich and the poor; and second that the poorest groups of the population have had an absolute decline in their real incomes (Ahluwalia, Carter and Chenery, 1979). Most of this evidence is based upon comparatively small groups over short periods.

An attempt has been made recently to estimate the number and proportion of the world's population with incomes less than $200 in 1950 and 1977. Using total Gross National Product data as the basis of the estimate, the proportion declined from 48.5 per cent to 25.5 per cent. However using consumption expenditure data, although the proportion fell from 53.4 per cent to 40.8 per cent, the absolute numbers rose from 1,297 million to 1,666 million (Berry, Bourguignon and Morrison, 1983).

The Sources of Food Output Growth: Expanding the Cropland

All the developing regions have experienced considerable increases in food output in the last 35 years, but the means by which this has been achieved has varied greatly.

Until the nineteenth century the principal means of increasing food output had been by increasing the area in crops. This was done by colonising land hitherto uncultivated or by reducing the part of the arable land left in fallow. In parts of Asia — and in Egypt — some arable land was multiple-cropped, with two staple crops being grown in a year. Since 1950 most of the extra food output in the developed countries has been

Table 10.6: Area in Major Food Crops[a], 1948–80 (million hectares)

	1948-52	1979-80	Change 1948-52 to 1979-80	(%)
North America	124.3	126.9	2.6	2.0
Europe	92.2	86.3	– 5.9	– 6.8
USSR	108.5	141.8	33.3	30.7
Oceania	6.6	17.2	10.6	160.6
Developed	331.6	372.2	40.6	12.2
Latin America	43.5	86.7	43.2	99.3
Asia	337.7	393.0	55.6	16.4
Africa	57.8	98.9	41.1	71.1
Developing	439.0	578.9	139.9	31.9
World	770.6	951.1	180.5	23.4

Notes: a. Includes all grains, potatoes, sweet potatoes, yams, pulses and oilseeds.
Sources: FAO, *Production Yearbook 1981*, vol. 35, Rome, pp. 93–137; *Production Yearbook 1976*, vol. 30, 1977, Rome, pp. 89–134; *Production Yearbook 1957*, vol. 11, 1958, Rome, pp. 31–2.

confined to the Virgin Lands scheme in the USSR in the 1950s, to an expansion of the wheat acreage in Australia and, in the United States, to a reduction of fallow and idle land in the 1970s. (Table 10.6). The area in food crops has increased more in the developing than the developed countries, 31.9 per cent compared with 12.2 per cent and in the developing regions the proportional increase has been greatest in Latin America, substantial in Africa and least in Asia, where however the absolute increase has been most. (Table 10.6). In Latin America most of the increase has come from colonising new land. There have been important expansions of the agricultural frontier in Brazil, particularly in the south of the country, and also northwards into the *cerrado* and the *selva*. There has also been movement downwards from the densely populated uplands of Central America and from the Andes, both to the Pacific coastlands and to the rainforests of the Caribbean coast and Amazon basin. In Africa there has been relatively little colonisation of new land, except in the Sudan, and much of the extra cropland has come from reducing the period in natural fallow in existing arable areas. In Asia much of the cultivable land was already in crops in 1950, and although there have been important additions, it has been a small proportionate increase. The existing arable land has however been used more intensively. In most parts of Asia temperatures are high enough to allow the cultivation of two cereal crops in one year if moisture is sufficient. The extension of the irrigated area (29 per cent of Asia's arable land is now irrigated) has allowed an increase in the intensity of cropping. Although multiple cropping indices are high in East Asia, they remain low in the less densely populated areas such as Thailand and Burma. (Table 10.7).

Table 10.7: Changes in the Intensity of Cropping in Asia, 1950–1980

	c. 1950	c. 1980
Taiwan	151	180[b]
Malaya[a]	101	160
China	130	150
Bangladesh	134	141
Philippines	126	136[d]
Pakistan	111	121
India	110	118
Nepal	125	117
Burma	107	111[c]
Thailand	—	101[e]

Notes: a. Riceland only
b. 1956–60
c. 1965–6
d. 1960
e. 1966
Sources: B.L.C. Johnson, *Development in South Asia,* Penguin, Harmondsworth, 1983, p. 62; D.S. Gibbons, R. de Koninck and Ibrahim Hasan, *Agricultural Modernization, Poverty and Inequality,* Gower London, 1980, p. 6; D. Dalrymple, *Survey of Multiple Cropping in Less Developed Nations,* United States Department of Agriculture, Washington, DC, 1971.

Although there have been major increases in the cropland in Latin America and Africa since 1950, the bulk of the increased world food output has come from the higher yields now obtained. Between 1950 and 1980 increased yields accounted for 85 per cent of the extra output of cereals — which are about four-fifths of the area in food crops. (Table 10.8). In the developed countries all but 3 per cent came from higher yields, but in the developing countries 40 per cent. However the proportion attributable to increased yields in the latter areas has increased from 18 per cent in the 1950s to 84 per cent in the 1970s.

Table 10.8: World Cereal Output, 1950–1980: Relative Contribution of Yield and Area to Increases in Output — Percentage

	1950–1960		1960–1970		1970–1980		1950–1980	
	Area	Yield	Area	Yield	Area	Yield	Area	Yield
Developed	−2	102	−5	105	42	58	3	97
Developing	82	18	25	75	16	84	40	60
World	56	44	12	88	25	75	15	85

Source: T.N. Barr, 'The world food situation and global grain prospects', *Science*, 1981, *214*, 1087–95.

Increasing Yields

Prior to the nineteenth century, crop yields increased very slowly in
Europe, except possibly in England and the Low Countries; but from
the early nineteenth century to the 1930s there was a steady upward trend.
However between the 1930s and the 1980s there has been a dramatic
increase, cereal yields doubling in North America and Europe since the
end of the Second World War. There have been equally dramatic increases in the developing countries. Indeed the average yield of all cereals
increased more in the developing than the developed countries between
1950 and 1980. In 1950 cereal yields were low in most developing countries, and lower than those in Europe, with the exception of East Asia.
Farming practices were simple. Little or no chemical fertilisers were
used, and in contrast to Europe little farmyard manure was used; indeed
livestock dung was rarely applied, and in India was often burned. Farming
implements were simple. In much of Asia wooden ploughs were drawn
by oxen or water buffalo, but in Africa the use of the plough was rare
outside the European-settled areas. In Latin America there were great
internal contrasts for tractors drew ploughs in southern Brazil, Argen-
tina and Uruguay, horses in Mexico, but nearly everywhere the hoe was
the major implement rather than the plough. From 1950 to the mid 1960s
yields were increased mainly by increasing labour inputs — the
agricultural population of the developing countries increased 50 per cent
from 1950 to 1980; land was weeded more frequently, irrigation systems
were extended and improved; and the seed bed was cultivated more
carefully. From the 1960s there have been substantial changes in farming methods in parts of Asia and Latin America. Improved varieties of
rice and wheat were developed in the Phillipines and Mexico and, independently, new high-yielding varieties of rice were bred in China.
These new varieties, when grown with irrigation, chemical fertilisers
and the use of pesticides gave substantial increases in yields over the
traditional varieties.
    The spread of the new varieties of wheat and rice was remarkably
rapid between 1965 and 1976–7 — there are no more recent estimates
(Table 10.9). However it should be borne in mind that the impact upon
food supplies is less dramatic than these data suggest, for wheat and rice
are not the only food crops — they are of very little importance in Africa,
and improvements in maize, sorghum and millets and the tropical roots
have been less successful, although the adoption of hybrid maizes has
made some progress. There have been considerable increases in fertiliser
consumption since 1950 though there is still a large gap between the

developed and developing countries, and also between Asia and Latin America, and Africa (Table 10.10).

Table 10.9: Percentage of Wheat and Rice Area sown with High-yielding Varieties, 1976–1977

	Rice	Wheat
China	80	25
Far East	72	30
Near East[a]	17	4
Africa	23	3
Latin America	41	13

Note: a. Includes North Africa,
Source: D.G. Dalrymple, *Development and spread of high yielding varieties of wheat and rice in the less developed nations,* Washington DC, 1978; R.C. Hsu, *Food for One Billion: China's agriculture since 1949,* Boulder, Colorado, 1982, p. 63; R. Barker, D.G. Sisler and B. Rose, 'Prospects for grain production', in R. Barker and R. Sinha (eds.), *The Chinese Agricultural Economy,* Croom Helm, London, 1982, pp. 163–81.

Table 10.10: Fertiliser Consumption, 1949–51 to 1980–81 (Kg. per Hectare of Arable, All Nutrients)

	1949–51	1980–1	1949–51 to 1980–1 (fold increase)
Developed countries	22.3	116	5.2
Latin America	3.1	46	14.8
Near East	2.4	34	14.2
Far East	1.6	38	23.8
Africa	0.4	10	25.0
Asian CPC	—	146	—
All Developing	1.4	49	35.0
World	12.4	80	6.5

Source: FAO, *The State of Food and Agriculture 1970,* Rome, 1971; *The State of Food and Agriculture 1982,* Rome, 1983.

Although the adoption of chemical fertilisers and new high-yielding varieties have been a major cause of increased yields in both the developed and developing countries, the improvement and extension of irrigation has been of great significance in the developing countries. In much of Africa and Asia seasonal drought and rainfall variability influence the choice of crops and the yield. Irrigation can extend cultivation into arid areas, ensure a reliable harvest in areas of high rainfall variability, and also make double cropping possible. In rice growing areas the plant must be grown with the stalk partially submerged for part of the growing

Figure 10.8: Contrasts in Agricultural Productivity, 1980–1982

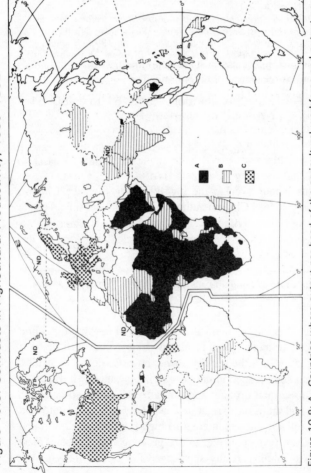

Figure 10.8: A. Countries where output per head of the agricultural workforce and output per
hectare are both 50 % or more below the world mean.
B. Countries where output per head of the agricultural workforce and output per
hectare are both 25% or more below the world mean; but not both 50% or
more below the world mean.
C. Countries where output per head of the agricultural workforce and output per
hectare are both 50% above the world mean.

season, and although two-thirds of non-Communist Asia's rice area relies upon rainfall or the floods of rivers, irrigated rice areas account for 60 per cent of the output. It is in these areas that the new high yielding rice and wheat varieties have been most rapidly adopted (FAO, 1979, p. 22). Although irrigation is a technique of great antiquity much of the world's irrigation is of recent origin. About 8 million hectares were irrigated in 1800, 40 million in 1900, and 120 million in 1950. By 1980 this had risen to 211 million. Only 4 per cent of the world's arable area is irrigated but in Asia 29 per cent is. Most of the increases in the last 30 years have come in Asia; in Africa irrigation is of little significance — although much needed — outside the Nile valley. (Brown and Eckholm, 1975; FAO, 1981; Gulhati, 1955)

In the developed countries the post-war period has seen a remarkable decline in the agricultural labour force, the widespread adoption of machinery and sharp increases in labour productivity. In the developing countries, in contrast, high rates of rural natural increase have been partially offset by rural-urban migration, but none the less the agricultural populations have increased by some 50 per cent, most rapidly in Africa and least rapidly in Asia. It is therefore debatable as to whether the use of machinery in agriculture is necessary or desirable, for labour is the factor of production that most developing countries have in abundance, and underemployment is widespread. Thus Chinese policy on mechanisation has varied, but the government has always admitted there is considerable surplus labour even in the highly labour-intensive Chinese agrarian economy. Indeed Chinese authorities fear that the spread of the Production Responsibility System may halve labour needs. (Hsu, 1982). Although the use of machinery is primarily intended to reduce the use of labour, often with increases in unemployment, it does contribute to the increase in food output, by allowing prompt and timely cultivation — critical at the end of the dry season in the African savannas — or by making double cropping possible.

Conclusions

Over the last 30 years world food output has exceeded the rate of population growth, as it has in all the developing regions except Africa. The proportion of the world's population suffering from undernutrition has diminished, and the numbers may also have fallen. None the less the problem of undernutrition is still formidable. The persistence of undernutrition is due to two causes. In much of the developing world — in countries containing two-thirds of the population — food supplies exceed national minimum requirements and inadequate diets are due to the poverty of the lower income groups. However in much of tropical Africa and South Asia, whilst poverty is a cause of hunger, the available food supplies — including imports — are insufficient to provide all with an adequate diet even if food supplies were allocated according to biological need rather than income. Hence in these countries there is a need not only for increases in income, but increases in food output. The shortage of food in these countries is due to many factors but basically to low productivity in agriculture (Cf. Figure 10.8 with Figures 10.1 and 10.4). Although no doubt it is correct to attribute the persistence of undernutrition to poverty it is clear that increased food output will be essential not only to supply the food deficits of these areas, but also to continue food output increases in other developing countries, for rapid population growth will continue well into the next century.

Note

1. FAO food production and food *per capita* indices are rebased at intervals and date from 1961–5. The longest series with a common base is 1961–5 to 1976; this is currently being extended to 1980.

References

Ahluwalia, M.S., Carter, N.G. and Chenery, H.B. (1979) 'Growth and poverty in developing countries', *Journal of Development Economics, 6*, 299–349

Bennett, M.K. (1941) 'International contrasts in food consumption', *Geographical Review, 31*, 365–76

Berry, A., Bourguignon, F. and Morrison, C. (1983) 'Changes in the distribution of income between 1950 and 1977', *Economic Journal, 93*, 331–50

Boyd Orr, J. (1937) *Food, Health and Income*, Macmillan, London

Boyd Orr, Sir John (1950) 'The food problem', *Scientific American, 183*, 11–15

Brown, L.R. and Eckholm, E. (1975) *By Bread Alone*, Pergamon, London

Food and Agriculture Organization (1952) *The Second World Food Survey*, Rome

Food and Agriculture Organization, World Health Organization (1973) *Energy and protein requirements: report of a joint FAO/WHO Ad Hoc Expert Committee*, Geneva

Food and Agriculture Organization (1977) *The Fourth World Food Survey*, Rome

Food and Agriculture Organization (1979) *The state of food and agriculture 1978*, Rome

Food and Agriculture Organization (1981) *Production Yearbook 1980, 34*, Rome

Food and Agricultural Organization (1983) *Production Yearbook 1982, 36*, Rome

Gopalan, C. (1975) 'Protein versus calories in the treatment of protein calorie malnutrition: metabolic and population studies in India' in R.E. Olsen (ed.) *Protein-calorie Malnutrition*, Academic Press, London

Grigg, D.B. (forthcoming) *The World Food Problem, 1950–1980*, Basil Blackwell, Oxford

Grigg, D.B. (1982) 'Counting the hungry: world patterns of undernutrition', *Tijdschrift voor Economische en Sociale Geografie, 73*, 66–79

Gulhati, N.D. (1955), *Irrigation in the World: a global review,* International Commission on Irrigation and Drainage, New Dehli

Gwatkins, D.R. (1980) 'Indications of change in developing country mortality trends. The end of an era', *Population Development Review, 6,* 614–44

Hsu, R.C. (1982) *Food for One Billion: China's agriculture since 1949,* Westview Press, Boulder, Colorado

Lardy, N.R. (1982) 'Food consumption in the People's Republic of China' in R. Barker and R. Sinha (eds.), *The Chinese Agricultural Economy,* Croom Helm, London, pp. 147–62

Le Gros Clark, F. and R.M. Titmus, (1939) *Our Food Problem and its Relation to Our National Defences,* Penguin, London

League of Nations (1936) *The Problems of Nutrition,* 3 vols., Geneva

Lipton, M. (1983) *Poverty, Undernutrition and Hunger,* World Bank Staff Working Papers, No. 597, Washington DC

Mata, L.J., Kranial, R.A., Urrutia, J.T. and Garcia, B. (1977) 'Effect of infection on food intake and the nutritional state: perspectives as viewed from the village', *American Journal of Clinical Nutrition, 30,* 1215–27

Morawetz, D. (1977) *Twenty Five Years of Economic Development 1950 to 1975,* John Hopkins University Press, Baltimore

Simon, J.L. (1980) 'Resources, population, environment: an oversupply of false bad news', *Science, 208,* 1431–7

NOTES ON CONTRIBUTORS

Dr J.W. Aitchison, Department of Geography, University College of Wales, Aberystwyth, Wales.

Dr I.R. Bowler, Department of Geography, University of Leicester, England.

Dr C.R. Bryant, Department of Geography, University of Waterloo, Ontario, Canada.

Mr P.J. Bryne, Department of Land Management and Development, University of Reading, England.

Dr G. Clark, Department of Geography, University of Lancaster, England.

Dr A.H. Dawson, Department of Geography, University of St Andrews, Scotland.

Professor D.B. Grigg, Department of Geography, University of Sheffield, England.

Dr B.W. Ilbery, Department of Geography, Lanchester Polytechnic, Coventry, England.

Dr M. Pacione, Department of Geography, University of Strathclyde, Glasgow, Scotland.

Dr W. Smith, Science Council of Canada, Ottawa, Canada.

Dr M.J. Troughton, Department of Geography, University of Western Ontario, London, Canada.

INDEX

abandoned land 17, 117, 175, 180
adoption rent 83
agrarian reform 6, 149–66
agribusinesses 5, 9, 25, 32, 99–102, 118, 124, 195, 227
agricultural censuses 46, 59, 175
Agricultural Circles 113
Agricultural Development and Advisory Service (ADAS) 205, 209
Agricultural Holdings Act (1984) 197, 202
Agricultural Holdings Bill 195
agricultural industrialisation 4–5, 24, 32, 93–121 *passim*, 234
Agricultural Mortgage Corporation (AMC) 205
agricultural policy 5, 6, 117, 124–48
agricultural regions 14, 16, 55, 60–2
agricultural revolution 4, 24, 94–7
agricultural vote 127
Agriculture (Amendment) Act (1984) 196
agrochemicals 75, 86
agro-food system 98, 104, 113
Agro-Industrial Complexes 104, 113
agro-industrial facilities 114
agro-industrial structures 119
agro-industrial technology 4, 102, 103, 115

Basal Metabolic Rate 240, 243, 244, 249, 250

Canadian Wheat Board 117
carrying capacity 95
Central Association of Agricultural Valuers (CAAV) 213, 217
classification 2, 3, 38–69
collective farms 98, 103, 109, 112–15 *passim*, 156–61 *passim*
collectivisation 7, 98, 104, 112–14 *passim*, 150–9 *passim*, 166
Comité des Organisations Professionnelles Agricoles (COPA) 128
Common Agricultural Policy 23, 133,

202, 216
compensatory allowances 136
Confagricultura 127
conservation 1, 20, 181, 188
consolidation 23, 141
contagious diffusion 73, 74, 75
contracts 10, 229, 230, 231
contract farming 86
contract integration 226, 229
conversion coefficients 45, 46
co-operatives 1, 10, 23–4, 86, 112, 117–20, 224, 232–3
corn-hog feeding 114
corporations 77, 78, 226, 227, 229
corporate farming 98, 99, 109, 116, 117, 225
cost-price squeeze 5, 116, 117, 131
Cottivatori Diretti 127
Country Landowners Association (CLA) 205
County Agricultural Boards 142
crop yields 258–61
cumulative diffusion 79

data orthonormalisation 61
decision-making 22, 25–31 *passim*, 76–7, 89, 126, 226, 228
deficiency payments 5, 133, 134, 135
differential adoption 85
direct marketing 23
disinvestment 179, 181
dormitory settlements 114

ecological optimum 1, 17
ecological stability 95
economic rent 16, 21
energy crisis 19, 20
enterprise types 51
equity goals 128, 133
estates 150, 152, 154, 198, 201
ethnic groups 22
European Economic Community 46, 49, 117–19 *passim*, 124, 128, 133–42 *passim*, 187, 216
eutrophication 105
expropriation 152